Pelican Books

Use and Abuse of Stat

W. J. Reichmann was born in Chelsea in 1920 and now lives in Acton. He has been a practising statistician in commerce and industry for some years, and this book is in part based on his experience as well as on some of his frustrations during these years.

He is a director of a number of companies with a wide variety of interests, and is a Fellow of the Royal Statistical Society, and a member of the Chartered Institute of Secretaries, the Market Research Society, the Mathematical Association, and other professional societies. He has edited an accountancy journal and contributed articles to other journals. His other books are *The Fascination of Numbers* (published in 1957), *Calculus Explained* (1964), and *The Spell of Mathematics* (1967), to be published in Pelicans.

W. J. Reichmann

Use and Abuse of Statistics

Penguin Books

Penguin Books Ltd, Harmondsworth,
Middlesex, England
Penguin Books Inc., 7110 Ambassador Road,
Baltimore, Maryland 21207, U.S.A.
Penguin Books Australia Ltd, Ringwood,
Victoria, Australia

First published by Methuen 1961
Published in Pelican Books 1964
Reprinted 1966, 1968, 1970, 1971

Made and printed in Great Britain by
Cox & Wyman Ltd, London, Reading and Fakenham
Set in Monotype Times

Contents

APPENDICES

Preface

Very few people nowadays can progress very far without at some point coming into contact with statistics. The general reader who wishes to understand what it is all about will not at first want to plough his way through a textbook even though, once his interest has been aroused, he may eventually come to realize the desirability of such an action. Before this, however, he will wish to be introduced to the subject in general terms. Statistics has something to say to him and the substance of what it has to say may be imparted without too much insistence being placed upon formal proofs and demonstrations.

There are many excellent statistical textbooks – both general and specialist – and no attempt has been made to emulate them here. A textbook is by its very nature primarily concerned with the mechanics of statistical method and the mathematical proofs of its theorems; there is usually very little space left for a general discussion of the position of the science and art of statistics with regard to the conditions in which it operates and of its relation to other branches of knowledge and inquiry with which it must cooperate.

This book is designed both for the general reader and for the student – as an introduction for the former and as supporting reading for the latter. Its object is not merely to describe how to calculate certain statistics but also to discuss how and in what circumstances they may be used and how they should not be used. Not less important is the emphasis laid throughout upon the dangers of misinterpretation and the difficulties which often beset the statistician's path of investigation.

The book does not demand any particular aptitude for mathematics although some basic mathematical demonstrations are of course necessary. These have been included in the main body of the work only where they are essential to the development; others have been relegated to the appendices but even there they have been kept to as simple a level as is possible, the object being to explain principles and not to solve particular problems.

Acknowledgements

Throughout this book references have been made to surveys, reports, Press articles, and other published and unpublished data. Where practicable, acknowledgements have been made in the relevant places in the text. Special acknowledgement is made to the following for their courtesy in giving permission for the reproduction of diagrams or substantial extracts from survey reports specifically referred to in the text: the Editor of *Lloyds Bank Review*; The Consumers' Association Ltd of 14 Buckingham Street, London; the Consumer Advisory Council of Orchard Street, London; the Institute of Traffic Engineers of Washington, D.C.; and the Director of Publications for Her Majesty's Stationery Office. The author also wishes to acknowledge the many helpful suggestions made by colleagues and friends.

1 The Age of Statistics

The Age of Statistics is upon us. Almost every aspect of natural phenomena and of human and other activity is now subjected to measurement in terms of statistics which are then interpreted, sometimes wisely, sometimes unwisely. Not even the more intimate details of human relationships have escaped the candid survey of the more relentless researchers and, as they probe ever more deeply and more widely into our affairs, it is perhaps not surprising that the layman should begin to wonder whether the statisticians are not getting a little beyond themselves. Generally, however, it is not the statisticians who commission or initiate such surveys; their function is to provide methods of interpreting the data and of ensuring that the accuracy of the results may be within specified limits. Some alleged surveys indeed pay little regard to statistical requirements at all and, as a consequence, are valueless. Nevertheless they masquerade as statistical surveys to the inevitable detriment of the reputation of the latter. The best that can be said for some surveys is that they are ill-advised in principle, but the function of the statistician is concerned with the mechanics of the surveys and not with the propriety of the surveys themselves.

There are two widely divergent views of statistics currently popular among the general public. One view is that published statistics are themselves invested with some quality of meaning not unlike the qualities ascribed to numbers by the Pythagoreans, and that they enjoy such a degree of infallibility that they may be accepted without question. This, of course, is just as nonsensical as the other and yet more popular belief that statistics can be made to prove anything and therefore, by implication, that in fact they can prove nothing. This is quite untrue and, although it is nevertheless nearer to the truth to say that statistics may be presented in such a way as to *appear* to prove anything, this is a very different matter altogether.

Both views are erroneous since they are based upon an ignorance or misunderstanding of the objects, scope, and discipline of true statistical theory and practice. But, whether correct or not, these general beliefs have in the past helped to build up a mass of antipathy which is even now considerable, despite the rapid advances and the contributions which statistical method has been able to make in the organization of our affairs and resources.

The statistician is often thought of as a man of another world as indeed are all mathematicians. But whereas the latter are contemplated with some awe by those whose own mathematics stopped at the multiplication tables, the particular class of statisticians is often regarded with some suspicion. This distinction is an odd one since a statistician is a mathematician yet, even if he is personally accepted as beyond reproach, the statistics in which he deals are themselves liable to be suspect. Why should this be so? Does some sinister influence come to bear upon the purity of arithmetic when sullied by its application to practical affairs in the form of statistics?

This air of suspicion derives, to a very great extent, from the activities of people other than statisticians. The practical value of statistics is cheapened and clouded by exaggerated press and advertising statements which are accorded a spurious appearance of statistical significance. These are often presented without any indication as to how the sets of data were collected, how they are to be interpreted correctly, and what true significance, if any, they possess at all. The reader is left to fend for himself.

There is a fairly clearly defined line separating statisticians from non-statisticians but this unfortunately does not prevent some of the latter from putting out masses of figures in the guise of statistics. It is they who, by accident of ignorance or design of dishonesty, tend to give a tarnished title to statistics generally in the opinion of non-mathematical citizens. But merely because some people habitually misuse grammar it does not follow that grammar is itself bad. Statistics may indeed be likened to a language. Linguists who speak a foreign language fluently can understand each other's conversation perfectly. In the same way no statistician can fool another statistician, although they may have legitimate differences of opinion. People who cannot speak French will candidly disclaim any ability in that regard; they do not understand the language and are quick to confess their ignorance. But, although they may not understand statistics, they will rarely be so ready with their disclaimers. This may perhaps stem from the fact that published statistics usually look so deceptively simple and disarming. Most people who can add two and two together tacitly believe that they understand statistics reasonably well enough.

When we are very young we are more likely to believe that a conjurer really is a magician. It is only as we become more experienced in the ways of the world that we come to know that he is nothing of the sort. The statistician is often regarded as something of a pseudo-magician. The statistician, however, cannot see into the future any more than anyone else can and, although he may help by projecting a shaft of light – however small – from his torch of knowledge, he is still acutely aware that his battery is all too often too weak to give anything but a shadowy outline. But the forecasting of quantitative outcomes of specified future events seems, to those who do not understand the principles employed, equivalent almost to clairvoyance. As a result they tend to belittle the statistician's efforts when they accuse him of failing to do something which he never claimed to be able to do.

Statistics as a science deserves better of its judges. Reputable statisticians do not indulge in the unwarranted manipulation of data nor do they stoop to other forms of chicanery in order to produce the answers they want rather than the true answers that are there awaiting discovery in the data. The problem, however, does exist for the general reader as to how he may judge the value of a statistical representation of facts placed before him. How can he tell whether the presentation is fair or whether he is being misled? Very often he will accept the presentation and all that it appears to signify or he will make some unjustified interpretation which is not supported by the basic data.

This is a very real problem and there is not always a simple solution. Nevertheless there are some elementary pointers to the truth and some criteria, to which reference will be made later, which may usefully be employed. The first step in the right direction, and therefore perhaps the most important, is the reader's realization that he must treat statistics objectively and that he should never accept them at their face value. The knowledge that statistics should be interrogated is clearly a prerequisite to the formulation of the questions which should be asked and is itself therefore a vital factor in the interpretation of data.

One of the reasons that the problem exists at all is that insufficient attention is paid to instruction in statistics. A mathematician, when confronted with a practical problem, will transform it symbolically into a form which is susceptible to mathematical treatment. That is, he will set up a mathematical model in the form of an equation or series of equations in varying forms by which he can translate the problems into the language of mathematics. Then, from the functioning of his model, he will relate the results obtained back to the original problem.

The art of teaching mathematics is in this translation of the problems into models and vice versa. Not all forms of mathematics can be treated in this way – abstract algebra is one which cannot – but applied statistics is a practical science and therefore can be treated in this way. Every new concept introduced has a metaphorical translation which, although not always providing an exact analogy, can nevertheless assist in putting the ideas across to the student. Statistics is a subject which, in its elementary stages particularly, lends itself to practical demonstration and, properly employed, it could well be integrated in the various stages of mathematical instruction. Integrated instruction is essential and due regard must be paid to the student's requirements. In the early stages of his instruction the practical applications are more important, because more easily understood, than formal theories. Professor M. H. A. Newman, in likening mathematics to a language,[1] has pointed out that children should be taught to talk and read a little before they begin on the grammar.

Mathematics has been defined by Bertrand Russell as the

1. Presidential address to the Mathematical Association, April 1959.

subject in which we never know what we are talking about;[2] or in other words that mathematics exists independently of the things it discusses. This view is contested by other mathematicians but even if it is true of pure mathematics it is certainly not true of statistics. Methodology as such may be divorced from reality but the interpretation of statistics by the use of those methods cannot exist independently of the outside world. To explain what is meant by a number, as distinct from its representative character, is a matter for philosophy but one does not need to understand the full significance of the concept of number in order to be able to use the numbers themselves. They may be used as the tools of mathematics, without the user worrying about their origins, in the same way as, in other applied sciences, he may use metal tools without considering the atomic structure of the tools themselves.

Provided one can set up a mathematical model which is truly or approximately analogous to a problem then mathematical method takes over and proceeds without regard to the problem itself until the solution is derived. The area of a circle is symbolically represented as πr^2 and this can be evaluated without regard to considerations as to what a square number (that is, r^2) really is.

The difficulty in statistics is in setting up the model, since there are often so many unknown quantities which need to be brought into some relationship between themselves and also some factors whose very existence may not even be suspected. Statistics deals with real things and therefore cannot escape from reality, but if it is possible to set up a sufficiently good model then the subsequent procedure is automatic. Data may be fed into a computer and the answer to some problem derived by purely mechanical processes without the computer ever needing to know what the problem is all about.

The elementary enumeration principles of statistics are by no means new. Their origins are to be found in the very beginning of mathematics. The basic concept is one of measurement and, as soon as man began to count his cattle and to cut notches in trees

2. *International Weekly*, Vol. 4 (1901).

to represent them numerically, so was born the science of statistics. But although the origins of statistics may be lost in antiquity much of its development is comparatively modern. As a science it is marching forward, probing, testing, and developing ever newer and more refined techniques, building sturdily upon itself and incorporating techniques from other branches of mathematics so that now it is assuming such proportions that even the statistician must specialize.

But statistics is not merely a science; the interpretation of statistics conforms also to the nature of an art even though much of it can be accomplished by strictly scientological methods. It is, of course, in the interpretation of statistics that the real value of the subject exists. There is no point in collecting data for the mere satisfaction of accumulation. Comparing measurements or other quantities, analysing effects of certain causes, tracing trends of changes in measurements against the backcloth of time or in varying conditions, deducing from these a norm of behaviour, and thence forecasting the probable outcomes of certain stated conditions – therein lies the value of statistics.

By the proper use and interpretation of data it is possible by such methods to produce meaningful results if, and only if, the basic conditions for the realization of those results are inherent in the data. Those who profess to find a significant result at any cost, although the available data do not support such a conclusion, perform a disservice to statistics as great as that resulting from the activities of others who, with set purpose, knowingly publish false or misleading presentations in a deliberate attempt to confuse uninformed opinion.

In the following pages an attempt has been made to help towards a better understanding of the scope and limitations of statistics and to indicate some criteria by which published or other statistics may be judged. The approach is at times necessarily negative since it is perhaps more instructive on occasions to explain what statistics do *not* show and thereby, by contrast, to accentuate the subject's more positive attributes.

There is a real gulf between the statistical and non-statistical worlds of ideas, and the statistician often finds it difficult to project his ideas across that gulf. This is perhaps partly his own fault in that his jargon, like all scientific terminology, tends to intensify the difficulties. It is paradoxical that even the more simple terms suffer the same defects, for statistics has borrowed a number of words in common use and has given them quite different or restricted meanings.

The mathematics of advanced statistical theory is strictly for the mathematician only and no attempt is made to reproduce it here. The language of elementary statistics, however, is relatively simple to master and, once learned, provides the key to a remarkable and fascinating subject. As Pollock[1] has said: ' A new language is a riddle before it is conquered, a power in the hand afterwards. . . .'

1. F. Pollock, *Clifford's Lectures and Essays*, Vol. 1 (1901).

2 Scope

The word 'statistics' has more than one meaning. Its use in the plural refers to descriptive statistics such as collected data, but in the singular it refers to statistical theory and method whereby the data are analysed. Thus the term is applied to the interpretation of a set of numbers as well as to the numbers themselves.

Of the latter little need be said provided that they are as accurate as possible, having regard to the type of measurement involved, and that they are relevant to a particular problem or investigation. In their recorded state within their own particular set the numbers may have little apparent value; their statistical meaning has to be drawn out of them by appropriate methods largely based upon the concept of comparison.

The concept of comparison is, of course, at the centre of much of human activity. The very measurements which provide the statistical data are themselves the results of comparison, the dimensions of the thing measured having been compared with a standard unit of measurement. Thus the width of this book may be measured by comparing it with standard units – that is, inches – on a ruler. Individual things, having first been compared with a standard unit, may then be compared with each other in terms of the standard unit, but it is essential that the standard should be a real one. Many people, for instance, suspect that during sale time the less reputable retailers first mark their prices up before marking them down. If a £3 article is reduced to £2 then there is a reduction of $33\frac{1}{3}\%$. If, however, the sale ticket shows a false original price of £4, then there is an apparent reduction of 50% and the sufferer from sales fever will regard it as being that much better as a bargain. This judgement results from the lack of a satisfactory standard unit – the article is a bargain only if it is worth more than the price asked for it irrespective of the former prices printed on the ticket.

Measurement is by definition impossible without comparison. In statistical method, however, comparison is not confined to the mere checking of measurements against some convenient yard-stick. The statistician does compare similar types of measurement over a period of time but he also compares their actual occurrences with the probability of their occurrences. The technique of quality control,[1] for example, enables managements to ascertain whether variations in measurements of apparently similar objects are due to chance or whether they are due to faults in the production line which, once identified, may be remedied.

This is an example of the development of statistics. Originally all statistics were historic; they were all concerned with the past and their processing was so protracted that the opportunity of using knowledge derived from them had often passed before the processing was complete. Now statistics is very much concerned with the present, and this important development has been made possible by the use of sampling techniques.[2] The statistician is now able to proceed beyond the comparison of actual measurements. He compares the measurements of those things actually measured and then by inference, again based upon probability theory, he projects the comparison to extend to all other things of the same kind.

This sampling method is an essential component of quality control techniques but it may also be applied in many other forms of statistical method. In quality control one is measuring something tangible and by inference accepting or rejecting the proposition that similar objects not measured will, within limits, have the same measurements. But sampling theory is not designed merely to detect the odd man out when he arrives as the herald of changing circumstances; in times of non-changing circumstances its use enables the statistician to estimate important values applicable to a whole population without having to measure every separate member of that population.

1. Chapter 20.

2. Chapter 16.

Information as to the heights and other measurements of a representative sample of men, for example, is of the greatest value to large tailoring firms dealing in ready-made suits since the measurements recorded may be applied to the allied measurements of the suits for men generally. This does not mean that they will necessarily manufacture garments having sets of dimensions in proportion to the data revealed or that every man will always be able to find a suit which will fit him perfectly. It is not the function of statistics to attempt the impossible but it is a function of statistics to provide information that may be applied to the possible. In this example the best arrangement would be one which would enable the manufacturers to satisfy most customers and at the same time avoid or reduce the need for alterations to garments.

The main function of statistics is thus to provide information. All sets of data give information of a sort; the degree of its usefulness will depend partly upon the purpose for which it is required and partly upon the manner of its collection. Tailors may obtain their information as to the more popular sizes of suits by the more direct method of noting which sizes normally sell most rapidly. This, in effect, is still a matter for probability calculations and is merely a different way of gathering the necessary information.

Masses of statistical data are gathered nowadays, particularly by government departments. Much of it may appear useless to the general reader, but it is not necessarily worthless to everybody merely because it holds no value for a specific individual. It may be valueless to him because he has no use for it, or it may appear to be valueless because he does not know how to use it, but others may have a very good use for it and be well aware how to use it to great effect. A surgeon will find a very good use for a scalpel whereas an artist would not – except, perhaps, in dealing with his critics – but scalpels are not useless merely because artists have no use for them.

Statistics therefore gives information and, in one form or another, is involved in every sphere of activity although it may not always be recognized. A person who compares the prices of tomatoes at different shops is mentally noting a primitive set of data to assist his decision as to where to make his purchase. This is a straightforward comparison of actual measurements. When he buys flowers before the Easter holiday because he knows that the prices are likely to be raised on Easter Saturday, he is comparing past measurements with probable future movements. When he runs his fingers through a pile of grass seed he is assessing the probable value of the whole pile from information derived from the sample which has passed over his fingers.

All these actions come so easily to him that he may not pause to realize that he is employing basic statistical ideas. These ideas are indeed inbred in every intelligent person and it is the universality of this fundamental application upon which rests the justification of statistics generally. Critics of statistics might well pause to reflect that they themselves often unwittingly employ its concepts and basic methods.

It is, of course, an implied condition that the information supplied by statistics shall be capable of being put to some valuable use. In circumstances of certainty the information may give exact details of happenings, but the application of statistics in its most useful stage is a process of providing information which will enable one to make decisions in uncertain conditions. This view of statistics as a decision-making process is of comparatively recent origin, yet it is implied in the condition that the information derived shall be useful. Any attempt to make decisions without the help of statistics is asking for trouble. It has been said that this is 'like asking a doctor to be responsible for getting a patient back to health but denying him the right to take his temperature, check the pulse, test the condition of his heart, or to ask questions about his past health'.[1]

Statistical method is essentially of the same structure as other forms of scientific method in which both inductive and deductive processes are employed. The scientist first makes his observations of relevant facts and, from a number of experiments, proceeds by a process of induction to the formulation of a theory which relates all or some of the experimental results to some common pattern. Having formed his theory he then proceeds by a process of

1. Lord Woolton, *Address to Royal Statistical Society*.

deduction from that theory to predict results of subsequent experiments which could not otherwise have been predicted. He then collects facts to verify or to deny the truth of his predictions and so the process continues, either by the development of the main theory or by the formulation of a new theory. This, of course, is a simplification of the actual process. Statistics is largely concerned with uncertainty and, where so much is uncertain, the results will often have but a limited application. Innumerable difficulties often confront the statistician in his efforts to pierce the mists of uncertainty and to light upon some significant fact.

Statistical method has in recent years made tremendous advances, demands having been made upon it at an ever increasing rate in business, politics, and in the sciences generally. Its success in assisting developments in these spheres has brought its own reward in an awakening interest in statistics and also in redoubled demands upon its resources. As a result of this, its problems become more difficult and varied. Its scope is continually widening and the statistician is being hard pressed to devise new or modified techniques to cope with the new demands. Many of the techniques are highly complex and sophisticated. Here we are concerned with the broader canvas of statistics generally and with the basic concepts upon which the theorist has been able to build.

Statistics is sometimes regarded as being remote from reality, but this view merely confuses the application of theories with the development of the theories themselves. A great deal of theorizing is always in progress. Just as statistics studies problems outside itself so also it must be continuously looking introspectively at itself. This abstract approach is as essential to the structure of statistical method as is the concrete approach to problems outside statistics. It merely means that one is studying the structure of statistical models rather than their use.

With so much research still being undertaken on statistical method itself, it should be apparent that statistical theory is by no means complete and will never have a satisfactory answer for every problem. There are indeed many problems still unsolved, just as there are many enigmas in other branches of science, but this does not detract from the usefulness of the solutions to those other problems which have not presented such difficulties.

As to the effective uses of statistics we may perhaps quote from a speech made by the Rt Hon. H. Macmillan:[1] 'To have any hope of carrying out their policies, the Government of the day must have knowledge of the facts as they are, together with such information as will help it to decide future trends.' It is true that he dryly remarked that it was equally essential for the Opposition to have these facts at their command in order to draw precisely opposite deductions, but the application of a wrong deduction does not overshadow the value to be derived from the application of the right one.

It is no exaggeration to claim that modern business in its present advanced stage of specialization would be impossible without the assistance of statistics. In conditions of mass production of goods it is uneconomic and sometimes impossible to inspect every single component produced. Even where 100% inspection is nominally effected, the human element will intrude itself to prevent it from being fully efficient. There is a world of difference between inspection as such and efficient inspection. It is impossible for wine-tasters to taste every drop of wine that reaches the shops for, quite apart from the physical impossibility of the task and the detrimental effects of making the attempt, such an action would, even if possible, defeat its own object by ensuring that no wine ever did reach the shops.

Statistics enters the production line long before the finished products are ready for sale. The modern manufacturer, being concerned in large-scale production which must be planned months or even years ahead, must keep one finger on the public pulse in order to ensure that he keeps his other fingers in the public purse. Consumer tastes vary from one period of time to another and the manufacturer must have the earliest possible warnings of probable quantitative changes in the demand for his products.

He will also need to know the nature of seasonal fluctuations in demand so that he can spread his production evenly over a full

1. Speech to Royal Statistical Society. (*Journal*, Pt IV, 1959.)

year. By knowing what stock he must have available at periods of peak demand he can arrange to build up this stock during the slack selling periods. Without adequate information he might well overstock and so tie up too much capital with no immediate opportunity of its producing revenue, or he might underproduce and thus be unable to meet the demands at the peak period.

Business men also need to keep a wary eye upon statistics of a more general nature. A number of different possible courses may be open to them and they will need to have an indication of which course will probably provide the best outcome. They must know something of general economic conditions, the amount of purchasing power available in a particular country, the total amount of hire-purchase debts outstanding, unemployment trends, changes in prices and supplies of commodities, as well as a long list of other statistics, not least of them bearing upon the likelihood of a particular political party being elected to a majority in Parliament.

Manufacturers must also know what their competitors are up to and must be able to assess the general level of trading and profit-making in their particular trade so as to measure their own performances. Ratios between figures published in their annual accounts provide pointers to the measure of success they are achieving in the utilization of their resources, and inter-firm comparison is made possible by this means. Only very large companies, however, at present disclose their total sales figures; most companies keep these figures well hidden because of the fear that they might reveal too much to their competitors. This attempt at secrecy is not always successful. For a particular trade and for a company of a particular size (judged by the capital employed, amounts owing by debtors etc.) it is possible to make a very good estimate of the total sales. A set of figures in the published reports will be consistent with the unpublished figures since together they comprise a complete and consistent set of accounts. Where perhaps six of these figures are published, it is not always difficult to estimate the seventh.

This is a very simple example of a statistical process which can make a set of data 'talk' and give information about related data. Certain sales turnover figures may also be derived in quite a different way by the process of sampling. A number of schemes exist whereby a sample of retail shops supply details of their sales, classified both by product group and by manufacturer. From these details it is possible to assess the proportionate share of national sales enjoyed by each main manufacturer. A manufacturer is thus able to express his own actual sales as a known percentage of the total market and, if he can obtain information as to his competitor's percentage share, he is able to estimate the actual sales represented by that percentage. This method is, of course, subject to sampling errors but the results achieved are sufficient for the purpose.

Statistics also helps business forward by its employment as a tool of industrial research and it is, of course, a most important factor in scientific research generally. Statistics assists not only in assessing the results of experiments but also in formulating the design of the experiments themselves. Methods have been evolved whereby more than one experiment may in fact be undertaken as one. The use of factorial designs and latin squares,[1] for example, enables the statistician to evaluate more than one factor within the structure of one series of experiments, and to remove effectively from experimental results such variability as is attributable to specific factors which are not themselves the primary subject of the experiment.

This is just one of the many ways which statistics has provided to make research more efficient and less costly. One of the worst drawbacks which statistics suffered for many years was that data collected on a large scale took so long to analyse. This has been partly overcome by sampling methods since a sample may be 'an instrument of policy, when the whole record may merely be a piece of economic history'.[2] It is, however, still desirable in some circumstances to take stock of a whole population (as, for example, with a national census) and the processing of the mass of collected data has now been speeded up to a remarkable extent by the use

1. Appendices 5 and 6.

2. Rt Hon. H. Macmillan. Speech to Royal Statistical Society. (*Journal*, Pt IV, 1959.)

of electronic computers. These can cope with calculations so rapidly that very few firms can find enough work for an installation of their own.

Computers have their limitations but these are at the lower rather than at the higher margin. There is a story of a representative who tried to sell a computer to a Chinese bank for he had learned that the clerks still used abaci. He agreed to a demonstration in which the computer and an abacus would compete and, in the event, the clerk had completed his calculation on the abacus before the problem had even been set up on the computer. This was because the calculations involved were sufficiently simple to be calculated on the abacus and were not sufficiently difficult to justify the use of a computer.

But, notwithstanding the use of computers, a great deal of statistical work is hard and painstaking, whether in using statistics as an aid to research or whether conducting research into statistical method itself. It is not, however, dull and unrewarding. Crowning success at sport or in the arts is not achieved without a great deal of patient practice and a great many disappointments beforehand. A properly conducted statistical investigation has all the spirit of the chase, amplified by the forward-pressing human curiosity in the search for knowledge, which in success brings great and rewarding satisfaction. Not all investigations take us to the end of the road, wherever that may be, but there is much to be learned on the way.

Statistics is essentially dynamic. It is continually moving forward and probing into new areas of application; more efficient techniques, greater computational facilities, and the resultant decreasing costs of investigation assist in the widening of its scope. In the widest sense, statistical success breeds its own rewards; the greater the national awareness of the ability of statistics to cope with the nation's manifold problems so the greater the demand for its services.

The future of statistics is assured since it runs parallel with human curiosity – both are limitless – but it must not be suggested that statistics is infallible. The science is by no means perfect. At the same time, it should not be held responsible for all the errors perpetrated in its name. There are some genuine paradoxes to be encountered in the subject, but there is also an abundance of fallacies – and fallacies result from the incorrect use of methods. Both the paradox and the fallacy provide traps for the unwary, but they may be avoided if recognized in good time. It is for this reason that they figure largely in the following pages.

3 Seeing Things

However well they may be tabulated, columns and rows of figures on a sheet of paper may convey little or nothing to anyone who is not accustomed to dealing with numbers. The mere effort of having to look at them at all may itself cause him to be worse confounded particularly if they are thrust at him by someone who evidently expects him to understand them at once. In circumstances such as these, one is apt to concentrate upon a particular section of the tabulation in the hope of deriving some significance from *that*, but therein lurks the danger of misplaced emphasis.

To be presented with a mass of unsummarized data is like being abandoned in the depths of a dense forest without the aid of a compass. Which way does one turn to find the way out; how can one tell east from west? A shaft of sunlight through the branches may lead to a clearing but it may still be leading you ever more deeply into the forest.

To be more easily understood, sets of data should be presented in as simple and straightforward a manner as is possible, having regard to the nature of the data and to the fact that some may be condensed and simplified much more easily than others. Business executives, faced with a financial loss, will want to know in which department the loss has occurred in order that they may call for a detailed report on that department. The first report therefore needs only to show the profit or loss for each separate department, perhaps sub-divided into product groups; it should not be cluttered up with data proving why the successful departments made a profit since this is not the object of the report.

The simpler the statistical representation so the more readily will it lend itself to interpretation. Data relating to a particular problem or state of affairs are collected in order to explain those specific conditions. Statistics is a tool of explanation and the value of the data will be diminished if they themselves require too much explanation. Simplicity of presentation is the corner-stone of the understanding of statistics.

Cold figures may be brought to life by illustrations. People find it easier and much less demanding to read pictures with a minimum of supporting fact – as is evidenced by the success of the illustrated newspapers. It is true that some of the illustrations in the latter are more of the nature of principal contributions, with facts as subordinates, rather than being incidental representations of the facts. Nevertheless the point is clearly established and accepted that diagrams and pictures help the human eye to convert data into food for the mind. A photograph will convey far more than can any written description of the object photographed. The truth of elementary algebraic equations, such as $(a + b)^2 = a^2 + 2ab + b^2$, can be grasped easily by the student if the equality is demonstrated geometrically.

Diagrams help enormously in the presentation of statistical data since they may be used to show at a glance the whole or central meaning of a set or some part of that set of data. Illustrations can give the dimensions of the whole and place the various factors in their right perspective so that the viewer may utilize his ability to discern differences in shapes, patterns, and physical magnitudes. Differences and similarities are emphasized. One only has to look to see them, and that is so much easier than having to think!

But even in the avowed desirability of simplicity one courts the danger of over-simplification. The simple things of life are often abused. It is hardly fair to blame this upon the simple things themselves for the blame should be laid squarely upon the abusers. Nevertheless, the innocent-looking criminal may be the most dangerous. It is possible to portray facts by means of charts or other diagrams in such a way as to give quite wrong impressions even though there may be nothing dishonest about the actual drawing. The reader, relying upon eyesight alone, is encouraged to draw the wrong conclusion. Photographs *can* lie; mirages *do* occur, and some of the most remarkable optical illusions are of the very simplest construction. But to know and to be able to recognize certain fallacies is half-way to defeating their effects.

Which segment of the line XYZ in figure 1 is the greater; is it XY or YZ?

Fig. 1

The ruler will show that XY and YZ are each of exactly the same length. Consider next the lines AE and BE in figure 2. How much longer is AE than is BE? Again it will be found that they are in fact equal.

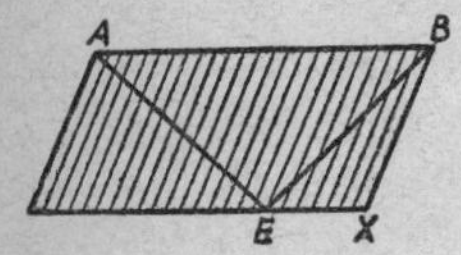

Fig. 2

The reader will now expect to find in figure 3 two lengths compared which are identical. And so he will, but no matter how long he looks at it, it is almost certain that his eyes will be very reluctant to accept the fact of equality.

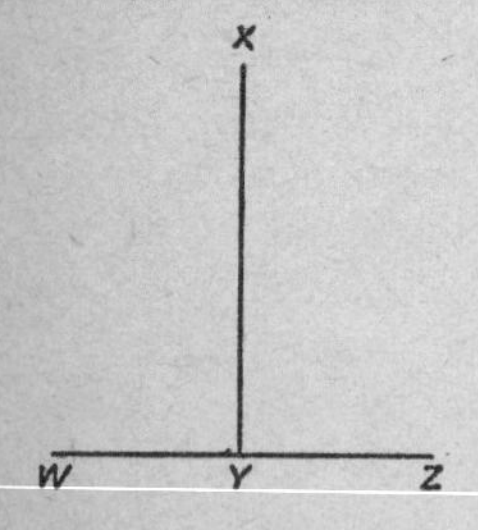

Fig. 3

It is clear that WY and YZ are equal, but XY appears to be considerably longer than WZ. Yet these two lines are also equal in length.

If such simple basic constructions can be so confusing to the eye, it is obvious that extreme care must be exercised when interpretations of more complicated construction are made since the same type of illusion may be present but carefully concealed.

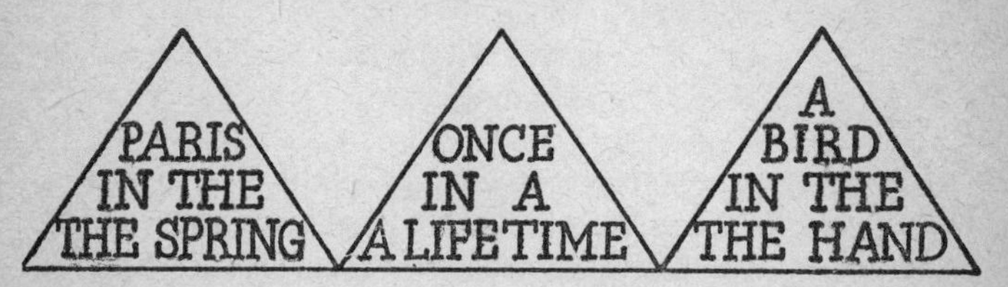

Fig. 4

Figure 4 incorporates three well-known phrases. The reader will have recognized them at once. Will he, however, have read them exactly as they are printed? That is most unlikely. In the first triangle he will probably have read 'PARIS IN THE SPRING' but another look will reveal that the word 'THE' is repeated. There is a similar repetition in each of the other triangles.

This demonstrates an additional point. Not only may diagrams be misleading in themselves; they may also appear to be misleading if one does not look at them properly. This illustrates at

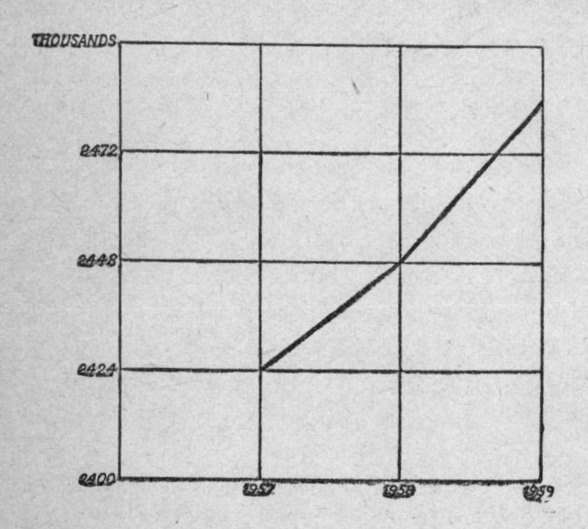

Fig. 5

once the absolute necessity for not taking things for granted and for not believing that one sees in a diagram that which one expects to see there without making really certain of the facts. Memory of past results similar to that apparently produced in a chart is sometimes apt to make the reader susceptible to suggestion and to see something which is not there. This is a common failing to which everyone is subject; the statistician and his reader must be constantly on their guard against it.

Statistical data may be represented by means of a number of different constructional methods, and optical illusions may occur or may be built into many of them. The most common method of construction is by line chart or curve (depending upon the nature of the data) and it is in the use of this method that many misrepresentations occur.

Figure 5 shows the trend of production of electric irons in Great Britain over the years 1957 to 1959. The figures show a slight rise in production. The reader who glances quickly at the chart may receive the impression that, since the production figure in 1959 is represented by a line three and a half times the length of the perpendicular representing production in 1957, then the production in 1959 was three and a half times what it was then. Yet this is far from being true, the actual figures being:[1]

1957	2,424 thousand (approx.)
1958	2,448 " "
1959	2,484 " "

The illusion arises because the chart is not complete. The zero line[2] has been suppressed and we are seeing only the top of a picture as if we are looking over a high wall that screens the lower half; it is impossible to appreciate a football match if all that one sees is the occasional glimpse of a football as it is kicked in the air. This suppression of the zero line is a weakness to be noted in many charts reproduced in the national Press where printing space is at a premium and where it is therefore not desired to reproduce the empty bottom half of the chart.

1. Board of Trade, *Monthly Digest of Statistics*, Central Statistical Office.
2. That is, the x axis.

The whole of a picture must be seen to enable one to understand its full value. A very different presentation of the data would be as in figure 6, in which it appears that the proportionate difference in production levels between 1957 and 1959 was almost negligible. Nevertheless there was a difference of some degree. This difference is grossly exaggerated in figure 5 but is almost 'lost' in figure 6. Some compromise scale between these extremes would be desirable.

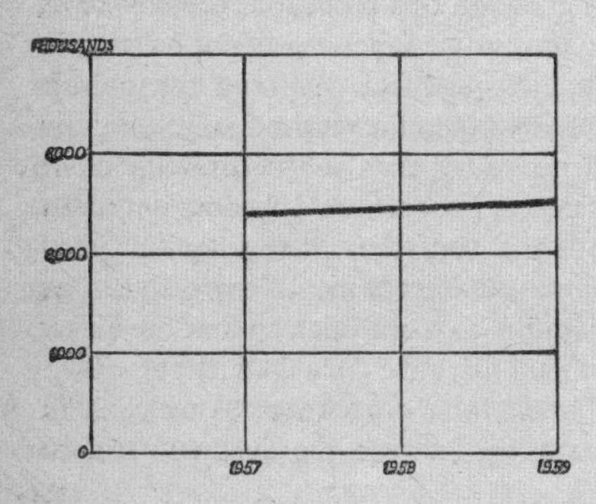

Fig. 6

It is always legitimate to cut out the blank bottom half of a chart, if the values charted are very high and the resulting line is always far removed from the zero line, provided that it is clearly shown that this has been done. The most useful way of doing this is to draw a jagged tear across the chart to show that there is a break in the scale and so to correct the otherwise misleading impression.

It should also be noted that in figure 5, since it showed only part of the picture, the scale of the chart was exaggerated to accentuate its apparent importance. The principle employed here is like that apparent in the action of the orderly officer who, when a soldier complained of his stew, sent him back for another helping of the same stew. If it is not good enough in single portions, then double the portions!

There are, therefore, two very good ways in which to judge the fairness of a line chart. The two questions to ask are: are we seeing the whole picture and is the scale exaggerated? The appearance of a chart is as much a function of the scale used as it is of the data charted. This inevitably raises the interesting point that even if the chart is honest and does not suppress the zero line, the gradient of the line shown may be increased or decreased by altering the scale. What yardstick is one to employ in judging whether the scale is exaggerated?

It is impossible to dogmatize on this point for the question is largely a matter of artistic proportions; yet it may not be right to draw a chart just as the draughtsman feels he would like to draw it. If only one variable is charted then the reader should carefully examine the scale and, if necessary, mentally re-draw the chart to suit his own personal sense of proportion. This does not mean that he can interfere with the truth but merely that he may represent the truth in some different way. If more than one variable is charted (i.e. by more than one line), so that the object of the chart is to compare their individual behaviour, the gradients of the lines will be altered in the same proportions by a change of scale but the visual relationship between the lines will still be altered.

The scale of a chart should be large enough to reveal significant differences or trends, since the eye cannot easily distinguish slight differences in gradients; on the other hand, it should not be so enlarged as to give improper emphasis to insignificant details. In a line chart (as distinct from a mathematical curve representing a functional relationship between variables) the lines themselves may mean nothing at all except as providing connexions between two or more points on the chart and to indicate what has happened between those points. The lines have no values in themselves.[1] The correct way to judge the relationship between values represented by two points is to measure the respective lengths of the *y* ordinates at those points. This is not possible if part of the chart is missing; in such cases the reader should give particular attention to the figures shown in the scale and should not let the expressed gradient delude him.

1. But see page 208 as to the distinction between discrete and continuous variables.

Just as the appearance of a chart depends upon the scale used, it is also a fact that the same pictorial shape has quite different degrees of significance according to its position on the chart in relation to the size of the values charted. In figure 7 the values of sales turnover and profit are represented by lines which,

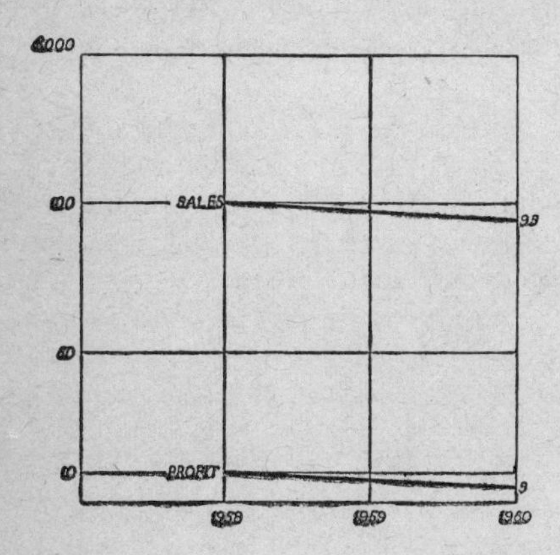

Fig. 7

between related points of time, are always parallel. Parallel lines, however, may have no virtue in themselves. Unless the lines are reasonably close on a sufficiently large scale, the significance of such lines, though geometrically similar, may be very different. They do, of course, show that the measurements of the two variables show similar absolute changes, but the changes in the measurements of each variable must also be related to that variable's own range of values. In 1958 the profit was equivalent to 10% of the sales turnover but it does not follow from the drawing of parallel lines that this percentage was maintained. In 1960 it had dropped to just over 5%. This is because the reduction of £5,000 had a much greater effect on a profit figure of £10,000 than it had on the larger turnover figure of £100,000.

Line charts, particularly those which portray time series, are subject to other misuses and these are referred to elsewhere. But optical illusions can occur in other types of diagram. One useful way in which data may sometimes be presented is by the pie method in which slices of a circular pie are cut in different sizes and in proportions which demonstrate the size relationship of individual data which, when taken together, sum to a total value represented by the whole pie.

Figure 8 is a pie chart showing the home population of the United Kingdom as at June 1959. It is possible to see at a glance

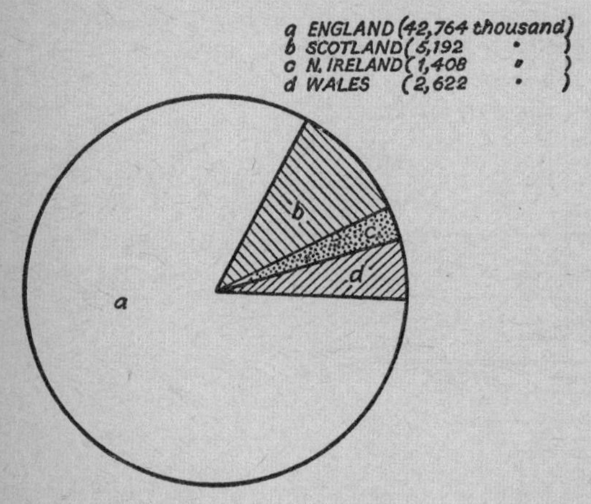

Fig. 8
United Kingdom de facto *home population* – (*June* 1959)
Total 51,986 *thousand*

that in size the population of England is much greater than all the other constituent quantities added together. This might be represented in another way by inscribing other circles inside the main circle so as to make a number of rings representing the various quantities. But this would not give a true comparison if the width of the rings varied in proportion to the sizes of the quantities to be represented.

It is the respective areas that must be in proportion but even when these are correctly drawn the result may still be misleading. Figure 9 is a simplified diagram of this kind and represents three quantities as (a) the outer shaded ring, (b) the inner unshaded ring, and (c) the shaded central circle. A glance at this diagram gives the impression that the order of magnitude of these quantities is; c, a, b. The area of the outer ring is obviously greater than that of the inner ring, while the central circle appears to be greater than both. In fact the area of this central circle is exactly the same as that of the outer ring.

Another useful method of representing data pictorially is by the histogram in which the quantities to be represented are

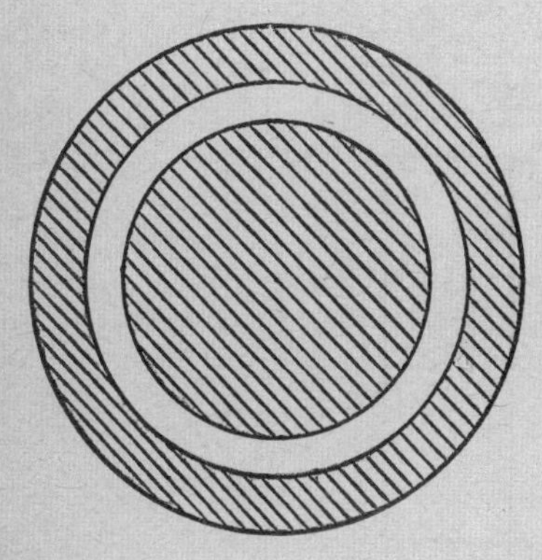

Fig. 9

indicated by the heights of contiguous columns. An example is given in figure 10*a*. Actually it is the area contained within a rectangular column which represents the quantity charted, but if the widths of all the columns are equal and constant then the heights of the columns are directly proportional to the quantities charted. If the widths of the columns differ, however, then the relative heights of the columns will give a misleading impression of the size relationship between the quantities represented. It is, in any case, difficult to obtain a visual assessment of the ratio between the areas of two columns unless they possess a common dimension.

In this kind of diagram it is again necessary that the whole diagram should be reproduced. If the zero line is suppressed, the appearance of figure 10*a* changes to that of 10*b* which gives quite a different immediate visual impression.

The histogram, although adequately serving its purpose, is too impersonal ever to become acceptable to or to make any real impact upon the ordinary citizen. In order to widen the appeal of published statistics it has become the practice to change the shape of the columns into symbols more closely associated with the nature of the variables represented. The histogram, which is a geometrical figure, becomes a pictogram consisting of a number

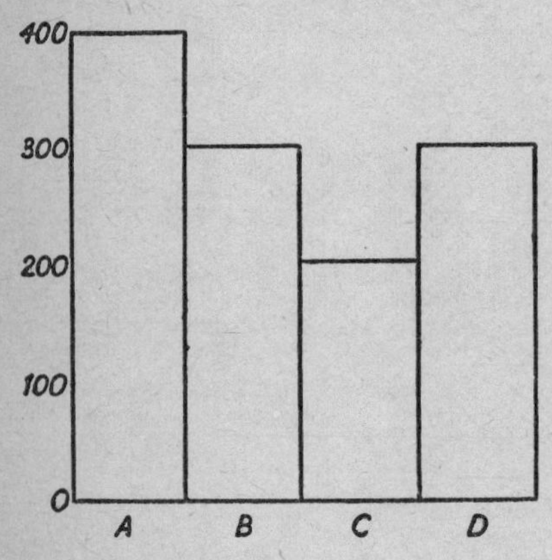

Fig. 10a

of pictures. If the columns of a histogram represent frequencies of men's heights then, in a pictogram, the columns might be replaced by pictures of men of differing sizes in proportion to the relative frequencies.

This sacrifices scientific presentation and it is doubtful whether it serves any really useful purpose. Those who are really interested in the subject will be able to derive their information from a histogram, whereas those who are not interested are not likely to pay more than passing reference to a pictogram. Even more important is the undoubted fact that this form of pictogram is more often misleading than not and is as regularly misunderstood. If the height of a man's picture is increased then the width must also be increased to the same degree so that the proportion of the picture may be maintained; otherwise the man will appear as an elongated freak. Thus, if the height is doubled the overall area covered by the picture of the man will be much more than doubled in the same way as when the side of a square is doubled the area of the square is quadrupled.

If it is desired to draw two pictures of men so that one is twice, or approximately twice, the size of the other, we are faced with the problem of constructing the pictures so that their size relationship is clear. It is, however, practically impossible to draw the

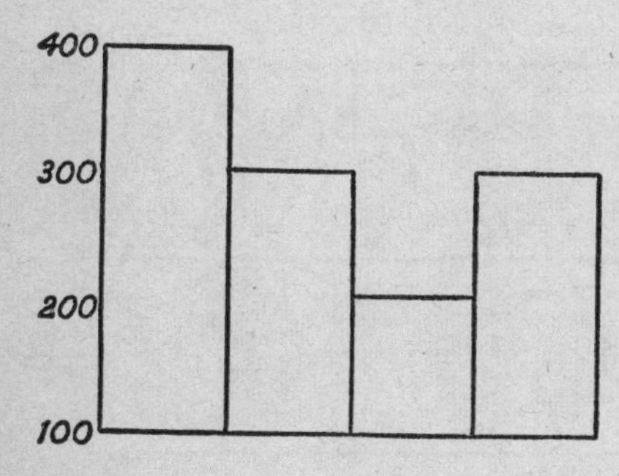

Fig. 10b

pictures in exact ratios and, since none of the separate dimensions (e.g. height or width) is in direct proportion to the overall ratio between the areas, it is very difficult for the untrained eye to see at a glance what that ratio really is.

Pictograms of this type sometimes have an odd effect on readers. There is the case, possibly apocryphal, of the egg farmer who was studying details of egg production. The compiler of the statistics, in order to emphasize the growth of annual total egg production as between the two years 1948 and 1958, had represented the annual figures by two differently sized eggs. These were drawn strictly to scale and showed a truly comparable size relationship. But the farmer was not to be deceived. 'That's ridiculous,' he claimed, 'eggs ain't no bigger now than they were in 1948!'

For these reasons other forms of pictogram have been developed in order to overcome some of the difficulties. Figure 11 is an example of how differences between the numbers of men employed on different duties may be emphasized.[1]

The little men are all shaded differently to help the eye to distinguish between the groups. More important than this, however, is the fact that the little men are all of the same size. It is easy to distinguish between numerical differences and similarities because these differences are represented as differences

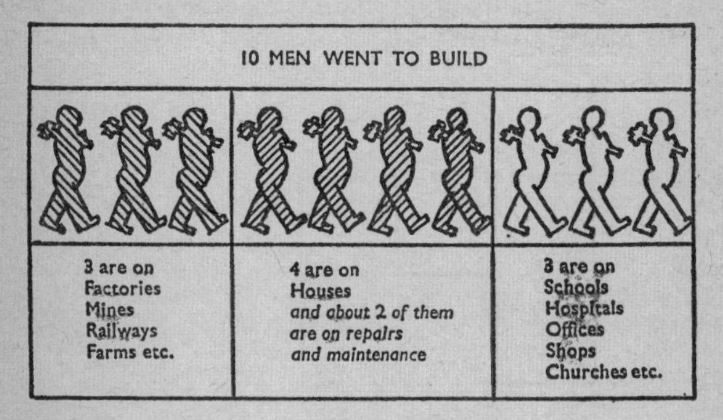

Fig. 11

Pictorial representation of proportion of workers in different sections of Building Industry (Survey '49, *H.M.S.O.*)

between numbers of little men instead of differences between the sizes of the men.

Pictograms take many different forms but the use of most of them involves the loss of important detail. They are intended primarily to attract the attention of the reader to the expression of a general basic fact (for example, that the standard of living has risen without giving precise details of the change) in the hope that he might be encouraged to read further into the detailed facts. Both histograms and pictograms have limited uses, the latter for their popular appeal, the former for basic knowledge on a more scientific level.

Some ingenious developments have made it possible to extend the use of histograms so as to depict more detail in one chart.

1. Reproduced from *Survey* '49 (H.M.S.O.).

Figure 12 is an excellent example[1] of a three-dimensional histogram in which it has been possible to show on one chart the percentage share of the world export markets enjoyed by four different countries over a number of years.

It is particularly noticeable that one rising 'staircase' (i.e. Germany) stands out more clearly than others. This is simply because the whole of every stair can be seen. In the others the

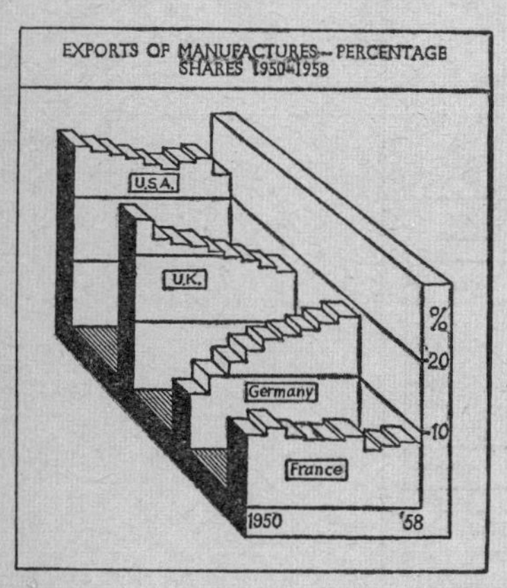

Fig. 12

Reproduced by kind permission of the Editor of Lloyds Bank Review

stairs are partially obscured by the sides of neighbouring staircases. This method would not be suitable if there were violent fluctuations in the levels recorded since parts of one staircase might be completely hidden behind another but it is ideal for its use in the example shown.

Use may also be made of the double histogram which enables two separate and distinct functions of the same variable to be included on one chart. Figure 13 shows details of personal incomes and taxation by income groups.[2] This sets out its information quite clearly and gives a very good visual impression of the facts.

1. *Lloyds Bank Review*, January 1960.
2. *Bulletin for Industry*, H.M.S.O., November 1959.

The last two examples demonstrate the value of pictorial presentation if it is effected with due regard to the limitations within which it must be confined and to the fairness of the resultant diagrams as representations of facts. The limitations of pictorial method are real ones and it is unfortunately true that

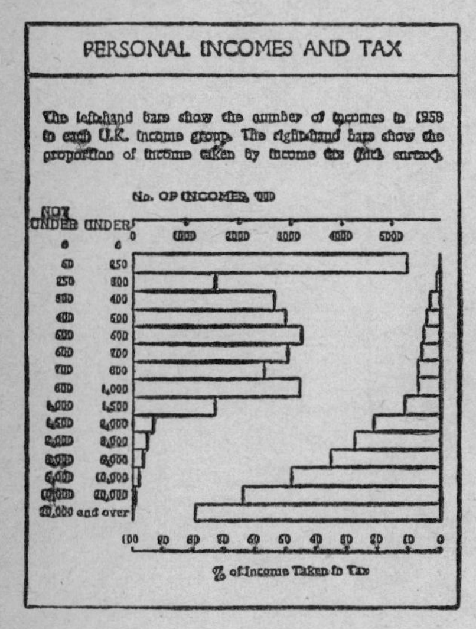

Fig. 13

they are often overstepped; but a knowledge of the nature of such limitations will help to distinguish the good diagrams from the bad ones. Diagrams should never be accepted without a close inspection of their *bona fides*. Things are very often *not* what they appear to be.

4 Reality

Pictorial illusions, as distinct from purely optical ones, may in fact project an illusory deduction wrongly interpreted from the data. Although a graphic representation (e.g. of the apparent correlation between two variables) may have assisted in suggesting a faulty deduction, the fault in an interpretation exists independently of its manner of expression. The misinterpretation may have one of several causes, ranging from a lack of logic in processing the data to actual inaccuracies and inconsistencies in the data themselves, and the statistician must follow a strict discipline of inspection throughout an investigation.

Before attempting to interpret data, it is first necessary to analyse them in such a way that it is crystal clear what they represent. Do the tabulations look as if they really contain the type of data for which the investigator is looking? Are they the latest available statistics and is the tabulation consistent within itself, or have the statistics altered since their original publication? In this connexion it is worth noting that publication can itself affect published data. There are, for instance, many procedural theories at large in the stock markets in which a certain percentage fall in the prices of equity shares is taken, in conjunction with other data, to point to the existence of circumstances in which the theorists believe they should sell their holdings. If this practice were widely adopted, the additional selling of shares would weaken the share prices still further so that the original price-falls would be seen to have borne within themselves the seeds of the further deterioration.

Published statistics are rarely in the exact form required for a particular investigation. Government departments and other public and private bodies which issue statistics cannot be expected to present them in just such a tabulation as may suit one investigation better than another. Instead they issue what, to them, seems to be the most important data tabulated in the simplest possible manner. Indeed, they very often have no idea of the purposes to which the data will be applied in the course of time – and they would sometimes be astounded if they did know!

The statistician must therefore rearrange the data that come to him and inspect them thoroughly in much the same way as a cabinet maker must ensure that he has the right timber and the right tools before commencing work. Just as the financial accounts provide the stepping-off place for an auditor's work, so do the data details provide the statistician with his basic requirements; neither document may be suitable in the form in which it is first presented but each provides a basis upon which one may work.

There is, of course, a limit to the possibilities of rearrangement. You can spell out the alphabet forwards or backwards but it is impossible to say it sideways! Similarly a photographic negative may be processed so that a man's face portrayed therein may be made to face either to the left or to the right in the final snapshot, but it is always the same side of his face that appears in the photograph. The other side of his face cannot be introduced by any manipulation of the negative since it was not photographed in the first place. There is a similar restriction upon the rearrangement of statistical data; they cannot be made to demonstrate any conclusion which is not contained within themselves.

In any event the rearrangement of data must be accomplished in such a way as to preclude manipulations which are deliberately or accidentally designed to produce the results which one wishes to obtain. This is not always so easy to achieve. Some children bear a strong resemblance to one of their parents, others do not; and yet friends of the family will often remark upon the similarity of a child to both parents. Are they merely trying to flatter the parents or are there optical illusions involved? Statistics offers a fertile field for illusions and it will have become apparent in the last chapter that pictorial representation cannot always be taken at its face value. It requires but a short extension of this thought to realize that mental illusion can be just as dangerous.

As a statistician Candide's mentor, Doctor Pangloss, would have been no more successful than he was as a philosopher. He would always have found the answer he wanted in his best of all possible worlds. The nub of a statistical problem is not 'What is the answer wanted?' but, instead, 'What is the answer wanted *for*?' The use to which data is to be applied will decide the most suitable method for their treatment.

The first questions to ask with regard to published statistics are who said so, what did he say, what did he *not* say, and what is his evidence? One is also sometimes tempted to ask whether he knew what he was talking about anyway! Absolute accuracy in subsequent calculations will be of no avail if the basic data are incorrect or unreliable in any way. It is important to know the origins of the data and also to know for whom and for what purpose they were accumulated. In times of short supply of goods, for instance, a retailer may order twice the amount of goods he requires so that, should his order be cut, he may still receive the amount he really requires. Anyone who added together all the increased orders from all retailers, to derive the total demand for a particular product, and who planned his production on that basis would conceivably end up in the bankuptcy court. The effective demand would be equivalent only to a half of his figure.

Particular care must be exercised when handling published condensed reports of surveys where only some isolated statistics are quoted or where the journalistic language has destroyed whatever precision was contained in the full report. The effect can often be to make nonsensical statements. In a statement on some aspects of a survey on the operating of staff canteens in factories[1] the following occurred: 'Out of every 100 employees in the London area who took a cooked meal, some 46 ate at a small canteen; 39 ate at a medium-sized canteen and 45 in a large one.' A simple addition will show that, although the statement starts off with 100 employees, it provides food for 130 and suggests that some of the employees were overdoing their calorie collecting by eating in two canteens at once!

1. *Canteen Management*, July 1960.

This is an error of presentation caused either by a misunderstanding of the basic data or merely by faulty phraseology – probably the latter. The number of diners quoted for each canteen is not meant to be related to 'every 100 employees in the London area', although that is what the statement says. The figure for each class of canteen is in fact related to 'every 100 employees employed by firms with that particular size of canteen', and that is a very different thing altogether. Thus, for firms enjoying the facilities of a small canteen it was found that 46 out of every 100 employees employed by those firms did in fact use the canteen – it is not that 46 out of every 100 in the London area used a small canteen.

In this particular instance the error is obvious since the reported figures sum to more than the assumed total, but a similar error might not have been so evident in differing circumstances. Condensed reports are more often than not useless for serious statistical work except perhaps for publishing the fact that a survey has been carried out. The statistician will never accept a condensed report but will always go back to the original data. It is a tragedy, however, that the abridged reports is all that the general reader is likely to see and it is scarcely surprising if he cannot understand it or construes it incorrectly.

The statistician is in a better position with regard to data for which he has himself arranged the collection. It is true that the collection of data is a routine matter and must be suited to the ability of the individual undertaking it. The data will therefore still have to be sorted for statistical purposes, but the great advantage over published statistics is that they will have been gathered for a specific purpose which was known before the collection of data commenced.

Adequate background information is absolutely essential particularly where definitions are involved.[1] The more questions one can ask about a set of data the better, since they may have been summarized in such a way as to exclude the particular information in which the reader is interested. One does not ask a serious question of another person unless one is first sure

1. Chapter 8.

that the latter is reasonably likely to be able to give a serious response. The ability of certain data to take part in an interpretation must be similarly tested.

In writing, many authors obtain ideas by a process of what, for want of a better term, is popularly called inspiration; but it is more often a result of consciously or unconsciously tuning one's mind into the right atmosphere of thought and of using a basic knowledge and drill to render oneself the more readily receptive of ideas. The understanding of statistics requires much the same kind of discipline. The statistician must have a receptive mind which can analyse the possibilities of a particular situation and he must make certain that it remains receptive. If you are expecting a friend to telephone you and he does not do so, it may be because your telephone is out of order; you could sit by the telephone for hours without receiving a message.

Statistics also requires a good grounding in experience. Data may provide a statement of past trends and possibly signposts to future ones. Not all the discoveries to be made may be on the main road; some unexpected discoveries may be found along the minor roads. Many roads are not signposted and it is therefore a great help to a stranger if he knows which road is probably the right one. Finding his way around is thus made easier but he must still take chances. There may be more than one way, however twisting, of reaching one's destination. But it is not always possible in statistics to know when one has reached the wrong destination. Real road signposts give a definite direction and quote a definite mileage figure. Statistical signposts, on the other hand, show a probable direction and quote probable values.

Thus, while it is usually open to us to learn from our mistakes, in some statistical investigations it is not always possible to identify the mistakes even when it is evident that mistakes have been made somewhere. To overcome this difficulty it is not only necessary to check the credentials of one set of data; it is also necessary to compare it with similar sets. Research in its early stages can never provide anything more than a guide. One scientific experiment alone would not be acceptable as proving a theory. Similarly the results of one survey alone should not be accepted without seeking corroboration from subsequent surveys or from independent sources. As De Morgan[1] has pointed out, an assessment of the guiding property of the loadstone, when first discovered, made it necessary to 'make many passages between ports that were well known before attempting a voyage of discovery'.

Before attempting to interpret data it is necessary to be sure that they are sufficient for and relative to the purpose and that they are as complete as possible. For this purpose one will usually need a reading of more than one event and more than one reading for each event. Absolute figures of one set may be of little practical use in themselves. If, for example, an examination is being made of data relating to products rejected by customers of a firm, the true significance of the numbers rejected cannot be appreciated unless it is also known how many items were originally supplied. Two customers may each reject three items. If the first customer originally bought 1,000 items, then he has retained 997 and the manufacturer will be quite happy about the general level of his products. If, however, the second customer originally bought only ten items, then the rejection of three is an indication that something may have gone seriously amiss. This one instance of a high rejection rate would not prove anything very definite as, in its isolation, it is not known whether it is representative of all goods supplied, but the manufacturer should take heed of the indication and should commence an earnest investigation.

It is not enough, however, that data should appear to be comparable; they must be shown to be comparable before comparison is attempted. An electric appliance may be returned under a guarantee for, perhaps, one of twenty different faults. It is not sufficient merely to calculate the total numbers of each type of fault discovered when the appliance is received back at the factory. Appliances with more than one fault will be enumerated in more than one category so that there will be a greater total number of faults than the number of appliances returned. These data, although necessary for some purposes (such as the comparison of the incidence rates of different types of fault), will not be enough. It is also necessary to know how many appliances had

1. *On the Study and Difficulties of Mathematics.* (Chicago, 1898) Chapter 1.

one, two, or three or more faults, since one fault may always be found closely associated with another. There may be a causal relationship between one fault and another; the latter may merely be a symptom of the first. To that extent they will not be strictly comparable.

The accuracy of the data should also be checked to ensure that there are not more appliances returned than there are faults recorded (unless the difference is accounted for by appliances being free from fault in spite of having been returned); that there are not more instances of one particular fault than there were appliances returned; and, where different models are included in the same tabulation, that no faults are recorded in respect of models which were not returned. This all seems so obvious that the reader may exclaim that surely that sort of thing does not happen. The author's experience, however, is that such things do happen; all three of these inconsistencies were discovered in one set of recorded data presented for interpretation.

It has been said that the actual evolution of mathematical theories 'proceeds by a process of induction – observation, comparison, classification, trial. . . .'[1] Statistics is not entirely a matter of mathematics but much the same stages of progress are required. Calculations based upon data may follow certain rigid mathematical rules and procedures and, because of this rigidity, may give a spurious appearance of exactitude to the results. But, although these calculations have formed part of a whole logical process of deduction, so that one follows naturally and necessarily from the other, the original data which is being processed is a collection of observations and the problem of finding whether they obey any form of law or order is originally a problem of induction. Mathematical treatment may help to reveal some form of order but the result cannot reveal something which is not already there. 'We cannot get more out of the mathematical mill than we put into it, though we may get it in a form infinitely more useful for our purpose.'[2]

1. E. W. Hobson. Presidential address to British Association for the Advancement of Science (1910). (*Nature*, Vol. 84.)

2. John Hopkinson (James Forrest Lecture, 1894).

Furthermore, statistics may reveal only a part of the truth. Just as one must take care not to produce spurious proofs from a set of data, so one must also be able to determine the limits within which any derived results may be applied. Certain factors may be immeasurable but their effect may be such that they cannot be ignored. Bare statistics may tell only half a story and it may well be that the unmeasured factors overrule any apparent result derived from the statistics. A Ministry of Transport report[1] on the operation of different types of road vehicle doubted whether centrally collected statistics could be fairly used as a basis for judgements about the efficiency with which large numbers of vehicles were operated.

> Such statistics cannot indicate how well the services which the vehicles provide suit the needs of the users. For example 'C' licence operations[2] are extremely varied, having clearly been adapted to the particular needs of individual firms at particular times and in respect of particular consignments in all manner of special circumstances. No statistics of the operation of the vehicles can measure the extent of the resulting economies and advantages which accrue to the users at other stages of the process of production and distribution.

Theoretically, a set of data can provide only one true answer to a specific question, for if there are two answers then the set of data contains inconsistencies within itself. The data may, however, be able to provide answers to more than one question even if the questions are so alike as not always to be clearly distinguished. This probably accounts for the fact that in some circumstances different people may draw different conclusions from the data. It is not *always* possible to be sure that one has the true result; different photographs may reveal different aspects of an object but it may be impossible to say which, if any, is *the* true representation.

Facts may be well hidden but no account of apparent logic and deduction can establish a result that is not contained in the data.

1. *The Transport of Goods by Road*, H.M.S.O., 1959.

2. A 'C' licence authorizes a trader to operate goods vehicles for the sole purpose of transporting his own products.

His theories must to some extent be influenced by his own human traits. Psychiatrists may stand as much in need of psychiatric treatment as do their patients. Doctors of medicine are certainly not immune from the human ills for which they prescribe. To some extent the application of statistics suffers from similar individual disqualifications. Since it deals very largely with human activity, interpretation must often be based upon psychological considerations.

But this disqualification may be minimized. Although each psychologist may be affected in this way, a group of them may possibly discern solutions to each other's problems since they, being themselves different individuals, will be affected in different ways. The pooling of their knowledge of each other, albeit coloured by their own realization of their selves, then builds up the science of psychology which is more than the total of the independent theories of its individual practitioners.

So it is with statistics. The statistician of today is building upon knowledge gained in the past by others and he is using techniques which are gradually becoming more and more efficient. Yet none of this will avail him anything unless he can get outside himself and view his problems objectively. Interpretation depends as much upon the interpreter as upon the data interpreted and he must be ever on his guard against unreality, prejudice, and illogicality. If, in the sometimes hectic atmosphere of modern business, he finds it difficult to maintain that cool detachment essential to his duties he might just as well, for all the good it will do him to keep pounding away at his problem, instead put on his hat and take a walk round the office block!

5 The Overworked Average

What is the average? Many people find refuge in this question when confronted by a mass of statistics. They are quite happy that they know all about averages and, in particular, they know that averages may be calculated quite simply without any fancy arithmetic. Everybody understands averages – until they come to use them!

It is an accepted fact that there tends to be a similarity of behaviour in many aspects of activity, human or otherwise, and in natural phenomena generally. One race of men may be taller than another. How do we know this? Certainly not because every man in race A is taller than every man in race B; there are bound to be some short A men as well as some relatively tall B men. But these exceptions will be few and it will be found that the heights of most men within each race will approximate to the average value for the whole of that race. Apart from the freaks, the measurements of men's heights will in each case tend to cluster round a central value, and it is this clustering which gives the average its meaning. There are so many readings clustered about the average as a central value that the latter may be taken as being representative of all the men considered as a race as a whole.

In choosing an average value for a set of data, the statistician is purposely selecting a particular value which shall possess this attribute of being representative of the set, so that the latter may usefully be compared with other similar sets. Surprise is often shown that it is possible to *choose* an average. Is there not just one average and that is the end of the matter? There is, of course, only one numerical value for any particular form of average of a set of data, but it is possible to choose between different types of average. The word *average*, which originated as a clearly defined insurance term, has taken to itself so many different shades of meaning that its use alone is now almost meaningless.

There are three main types of average – the mean, the mode, and the median; and the first type may be subdivided into arithmetic mean, geometric mean, and harmonic mean. These are all different types of representative value and, being recognized as such, the noun *average* is rarely used in statistical work when referring to actual values. The statistician, when mentioning these values, will always refer to them by their specific names. If he wishes to talk about them collectively he will refer to them as measures of central tendency rather than as averages. But elsewhere they are often referred to as averages without any indication of the type of measure being used and, since each type is derived quite independently of the others and has quite a different meaning, it is not surprising that this sometimes causes misunderstanding.

Nevertheless the use of the word 'average' as a generic term is permissible where in its context it is not referring to a specific value or where it is clear that it is not synonymous with any particular kind of the different types of average employed.

The three different types of mean are intended for different types of problem, in the same way as different tools are required for different jobs. The means differ from the other measures in that they are results of calculations involving individual values in the data sets. The median and the mode, however, are specific observed values within the set and are not derived from mathematical calculations.

The simplest form of mean is the *arithmetic mean*. It is the easiest to calculate. One just adds up all the values, divides by the total number of values, and the result is the arithmetic mean. Expressed mathematically

$$\text{Arithmetic mean} = \frac{S}{N}$$

where S is the sum of all the values, and N is the number of values. This measure is the most popular one in everyday use and has many applications. Sport abounds in examples of this type of measure, as also do statistics which are produced in business and commerce generally. It is so simple and foolproof! Well, the calculations may be so, but the application of the result may very well not be so foolproof. An arithmetic mean may not merely lack significance; it may be positively misleading.

Here is a simple form of business take-over bid. There are two greengrocers in a market. Dealer A regularly sells new potatoes at two pounds for a shilling, while B sells a slightly lower quality at three pounds for a shilling. Each of them sells 60 lb. each day; A therefore receives 30 shillings, while B receives only 20 shillings. One day B buys up A's business. He decides to mix the two kinds of potatoes, but also decides that he must treat his customers fairly. Previously, by going to each greengrocer the customers could have bought 5 lb. for two shillings (that is 2 lb. and 3 lb. respectively from A and B). Very well then; so he decides that he will maintain this average price, equivalent to $2\frac{1}{2}$ lb. for one shilling and at this attractive price he sells 120 lb. He has therefore sold the same total amount as was previously accounted for by the combined amounts sold by A and B and he mentally clocks 50 shillings in his till, this being the total of the individual amounts previously received by A and B (i.e. 30 + 20). When he counts his takings, however, he finds that he has only 48 shillings. What has happened to the other 2 shillings? Should he sack his boy assistant?

Fortunately he decides to sit down and work it out on paper. He proves that his boy is honest but he still cannot understand why the difference arises:

(i) Before take-over:	60 lb. @ 2 lb. for 1s. =	30 shillings
	60 lb. @ 3 lb. for 1s. =	20 "
	120 lb.	50 "
(ii) After take-over:	120 lb. @ 5 lb. for 2s. =	48 "

That's silly, he claims. A customer could have bought 5 lb. for two shillings, before we merged the business, by buying 3 from me and 2 from A. Why cannot I sell them at the same rate and obtain the same income? The fallacy of this reasoning was that,

mean value. In other words the mean does provide a basis whereby abnormal values may be recognized.

The mean is representative of set (a) because it is the result of averaging similar quantities; that is, within a reasonably small range. The mean of set (c) is not representative of the whole set because it is the result of averaging dissimilar quantities. The arithmetical calculations are correct in each case but the resulting mean of set (c) is meaningless as a representative value. Similarly only the measurements of things with a common identity should be averaged. It is not easy to define exactly what constitutes common identity in this context and it is really a matter of common sense. Tigers and domestic cats are both members of the cat family, but it would be ridiculous to compare statistics of one with the other or to aggregate the two sets of data, calculate the mean and then claim that the result constituted a statistical mean for the cat family in general.

In the same way it is just as meaningless to quote the value £500 as the mean income of three individuals whose separate earnings are £1,000, £300 and £200. The earners are clearly in different classes and the mean quoted would appear to place them all in another quite different class. It is misleading in such circumstances to consider income as one indivisible subject for statistical analysis since it has too wide a connotation. Much the same principle is involved in averaging the shareholdings of the members of a limited company where there is a wide difference between the sizes of some of the holdings. A company whose share capital is £2,000 and which has 100 members would have a mean holding of £20, but the real facts could be as follows:

1 shareholder holds	1,010 shares		=	£1,010
99 shareholders hold	10 ,,	each	=	990
100				£2,000

To quote the mean value of £20 is to suggest that the capital of the company is mainly held by small holders. The truth, however, is that one member holds more than 50% of the capital and therefore has control of the company. Furthermore the calculated mean is twice as great as the individual holdings of 99% of all the members!

It is therefore a most important rule that only like objects should be included in mean calculations so as to ensure that there is some approximate relation between the mean and all the individual values. Similarly, when comparing the means of separate sets of data, it is first necessary to ensure that the sets are themselves comparable before attempting to compare statistics derived from them. Sets of data may or may not be comparable depending upon the kind of comparison desired. They may be comparable in one way although not in another, and it is not sufficient that they should be nominally comparable. The mean age of all the residents in one street on 1 January was 63 years. On 1 February, it was 60 years, and a direct comparison might suggest that the inhabitants were getting younger instead of being a month older! All that had happened, however, was that the oldest inhabitant – who was 90 years old – had moved to another street. The composition of the street residents was therefore not the same in each set of data and the comparison does not compare identical classes. Although all the members of the second class were included in the other, the two classes were quite distinct.

For this and other reasons the mean should never be accepted as significant without supporting credentials. The following table shows the trading results of a company by separate departments over two different years:

	Profits (+) and losses (–) £000	
Department	*1959*	*1960*
A	+30	+15
B	+20	–20
C	–15	+ 1
D	+40	+60
E	–10	+ 9
Totals	+65	+65
Mean	+13	+13

If one accepts the two yearly means as showing that no changes have occurred between the two years, there is a shock awaiting anyone who looks back at the departmental details from which the means were derived. There has been a shifting of profitability as between departments and the real position is very different; yet comparison of the two means hides the change entirely.

In calculating the mean of a set of data which is expressed as frequencies within classes, so that individual values are not stated, it is usual to assume that the mean within each class coincides with the mid-mark of the class. Thus, if a class is expressed as including measurements between 1,001 and 1,500, the mid-mark will be 1,250. This value is assumed to be equivalent to the class mean and is included as such in the calculation of the set mean. This is not always reliable, particularly where all the measurements tend to cluster around the class margins and where the classes are too widely framed.

Consider the following:

Salary class (£)	*Frequency*	*Mid-mark*	*Mid-mark × frequency*
501–1,000	5	750	3,750
1,001–1,500	20	1,250	25,000
1,501–2,000	20	1,750	35,000
2,001–2,500	5	2,250	11,250
Totals	50		75,000
			Mean = 1,500

The actual individual salaries, represented within classes as above, could be composed as follows:

Salary (£)	*Frequency*	*Salary × frequency*
900	5	4,500
1,400	20	28,000
1,800	20	36,000
2,300	5	11,500
Totals	50	80,000
		Mean = 1,600

The latter mean is the correct one for the set as a whole, so that the mean derived from the class frequency calculation is about 6% short of the correct value. This is because the mid-mark in each class is not representative of the class. It is a paradox in this form of calculation that an overall percentage increase in all the salaries could reverse the position. A 15% increase on all salaries would raise them to:

Salary (£)	*Frequency*	*Salary × frequency*
1,035	5	5,175
1,610	20	32,200
2,070	20	41,400
2,645	5	13,225
Totals	50	92,000
		Mean = 1,840

Represented in classes, these data would be transformed to:

Salary class (£)	*Frequency*	*Mid-mark*	*Mid-mark × frequency*
1,001–1,500	5	1,250	6,250
1,501–2,000	20	1,750	35,000
2,001–2,500	20	2,250	45,000
2,501–3,000	5	2,750	13,750
Totals	50		100,000
			Mean = 2,000

This time the incorrect value is higher than the correct mean because each mid-mark is greater than the actual salaries within each class. It should also be noted that, after the 15% increase in salaries, the correct mean value is also 15% higher, as one would expect, whereas the incorrect value after the increase (£2,000) represents an increase of 33⅓% on the original incorrect value (£1,500). This divergence arises because of the transfer of the frequencies to higher classes and, whereas in the original classes the salaries were close to the upper class-limits, the increases are sufficient only to take the frequencies just in to the higher classes so that they are closer to the lower class-limits.

This example is an extreme case but it helps to emphasize the further important point that, where an individual salary is close to the upper limit of one class, only a very small increase in salary might lift the individual into a higher class without making much material difference to his position. If classes are framed as (a) up to £500, (b) £501–1,000, (c) £1,001–1,500 etc, then a man earning £999 is in class (b). He only needs an increase of ·002% to lift his salary to £1,001 and so to place himself in class (c). An increase of £2 is all that is necessary to effect this translation to a higher classification; it is not necessarily true that individuals in one class are materially better off than those in a lower class.

This is of some importance since salary ranges are widely but perhaps erroneously accepted as indicating social standing and ways of living. This stems from the fact that although the classification of values by frequencies within classes was originally merely a statistical convention to facilitate calculations, the classification of numerical strata has come to be regarded as being invested with an intrinsic significance of its own. Transfers between classes often give false impressions as to the significance of differences between values, in the same way as a shopper, who will not pay twenty shillings for an article but will permit herself to be persuaded to pay nineteen shillings and elevenpence halfpenny, confidently deludes herself that she has no equal in the financial world.

The arithmetic mean cannot properly be employed for all sets of data. It cannot, for example, be used to average rates of growth or rates of speed. A man cycles to a point one mile away at 2 m.p.h. and returns over the same distance at 6 m.p.h. He therefore averages 4 m.p.h. – or does he? He does not. It took him 30 minutes to cycle the first mile and 10 minutes to return, so that in fact he cycled 2 miles in 40 minutes and his average speed is therefore 3 m.p.h.

Confusion over the calculating of average speeds often gives rise to misconceptions. A magazine contributor stated in an article[1] that to average 50 m.p.h., having covered 5 miles at only 30 m.p.h., would mean that a motorist would have to cover another 5 miles at 70 m.p.h. In point of fact the second run of 5 miles would have to be covered at 150 m.p.h. in order to achieve the desired average speed of 50 m.p.h. over the full 10 miles.

This problem can be looked at in another way. If in the first example the original average of 4 m.p.h. is correct the problem may be posed in this way: if a man cycles one mile at 2 m.p.h., at what speed must he cycle on his return journey in order to average 4 m.p.h.? At 2 m.p.h. he will do the first mile in 30 minutes. If he is to achieve an average of 4 m.p.h. for the full journey then he must cover the full two miles in 30 minutes. But it takes him 30 minutes to cover the first mile and he must therefore cover the second mile in 0 minutes! The average of 4 m.p.h. is therefore impossible to achieve so that, no matter what speed he achieves on his second mile, he can *never* achieve an average of 4 m.p.h.

The calculation of 4 m.p.h. was derived by taking the arithmetic mean of the two speeds of 2 and 6 m.p.h. respectively. The result is incorrect because this in effect averages distance related to time, but the time element is different in each case. The correct result may, however, be achieved by averaging the time related to distance because the distance is the same in each case. The only way in which the distances may be averaged is by weighting them so as to equalize the time elements. A more direct method, however, may be achieved by calculating the *harmonic mean*, which is calculated by dividing the number of values by the sum of the reciprocals of the individual values. The correct average speed may therefore be calculated as being equal to:

$$\frac{2}{\frac{1}{6}+\frac{1}{2}} = \frac{2}{\frac{2}{3}} = 3 \text{ m.p.h.}$$

The harmonic mean is always the right one to use to average rates and prices. The greengrocer's calculation[2] to find his

1. *Safety Fast* (September 1960).

2. Page 56.

average selling rate would have been:

$$\frac{2}{\frac{1}{2}+\frac{1}{3}} = \frac{2}{\frac{5}{6}} = \frac{12}{5} = 2\tfrac{2}{5} \text{ lb. for 1s.}$$

The harmonic mean deals with rates which are not dependent upon each other. The *geometric mean*, however, is used to average rates of growth where a subsequent measurement (e.g. national population statistics) is dependent upon previous measurements – the growth building up upon itself so that the rate of growth is continually increasing. The population of a country, for example, increases proportionately to the number of inhabitants provided there are no complicating factors introduced by emigration or changing birth and death rates.

As the population grows there will be more births which will again increase population and lead in time to yet more births. It was this problem which worried Malthus[1] so much; he visualized the population outstripping the sources of food and concluded that famine would eventually sweep the world to counteract the growth if wars had not already done so.

If a population was 2 million in 1940 but had grown to 4 million in 1960, it will clearly have doubled itself in twenty years. What would the population figure have been in 1950 after only ten years? It would not have been 3 million (which is the arithmetic mean of 2 and 4 million) since the increase in population would have been more rapid in the later years of the period than it would have been in the earlier years. The real figure would have been approximately 2·8 million. This estimate is calculated as the geometric mean.

The geometric mean, like the other means, is the result of a mathematical process – but it is not the same process. The geometric mean is calculated by multiplying together all the values in a set and then extracting the *n*th root, where *n* equals the number of values. Thus the square root is required to obtain the geometric mean of the two population figures:

$$\sqrt{2\text{m.} \times 4\text{m.}} = \sqrt{8\text{m.}^2} = 2{\cdot}8 \text{ million (approx.)}$$

1. T. Malthus, *Essay on the Principles of Population*, 1798.

These, then, are the three types of mean – the arithmetic, the harmonic, and the geometric – all of which are derived by the use of simple mathematical principles. They all have one aspect which at first glance appears to be a disadvantage. Although all the values in a set may be whole numbers, it is in the nature of things that any of these means will often be calculated as fractional equivalents which do not correspond to any of the actual values. This is a disadvantage which may be more apparent than real. If the variable is continuous, the circumstances do not exist, but it does lead to some superficially odd results if the variable is discrete – as, for example, when the mean household size is shown as 3·23 persons.[1]

Similarly a cricketer's average may be reported in the sporting columns as 80·3 runs per innings, yet the batsman can never achieve 0·3 of a run and will be run out if he attempts it! It is worth noting that these cricket averages are not true means. To obtain the arithmetic mean one would divide the total number of runs by the number of innings. But his 'average' is intended to mark the cricketer's prowess as a batsman. If any of his innings is not completed through no fault of his own (that is, because he has survived all his team mates) it is assumed that he would have been able to score more runs if the innings had not ended. A convention is therefore adopted whereby the number of runs is divided by the number of *completed* innings. Nevertheless, all the runs are included even though some were scored in uncompleted innings.

This naturally gives an inflated value to the 'averages' of those batsmen who manage to outlast the rest of their teams. One wonders what would happen if, by some freak of luck, a batsman managed to pass through the whole cricket season without once losing his wicket. What would his average be? Would it be the total of runs he had scored or would an attempt be made to subject this total to the impossible division by nought – this being the number of completed innings? No doubt someone would think of a new convention at the crucial moment.

The other measures of central tendency – the mode and the

1. National Food Survey – 1956 (*Annual Report* published 1958).

median – are not subject to these mathematical quirks, for they are not normally derived mathematically at all. Although there are some exceptions to the rule, these measures are normally actual values which by observation are selected as being representative of their set.

The *mode* is the value which occurs most frequently. Its main importance is in indicating the value of a substantial part of a set of data and it is most useful for interpretative purposes when most of the values tend to cluster around the modal value. Although the latter is that which occurs most frequently it does not follow that its frequency represents a majority out of all the total number of frequencies. This, in a sense, is analogous to the outcome of three-cornered election contests. The victor, although he has been accorded more votes than either of his rivals, may nevertheless have been elected by a number of votes which is lower than the combined votes of the others. It might be said indeed that there were more votes against him than there were for him, although the mood of voters in such cases is unpredictable and estimates cannot always be made as to who would have received most votes if there had been only two candidates. But this doubt, whether measurable or not, is one which cannot be ignored; in the same way the mode by itself could prove misleading in certain cases.

The *median* of a set of values is normally the central value in terms of magnitude. Thus if there are seven values the median is equivalent to the fourth value, there being three values below and three values above it. It is, of course, important that, before a set is divided in this way, the values must first be ranged in order of magnitude on the military tallest on the right shortest on the left drill formula. Where there is an odd number of values in a set and all the values are whole numbers, the median will also be a whole number. Where there is an even number of values, there is not a central value in the same sense. Instead, there is a central pair of values and the median is calculated as lying half-way between these two values.

The object of a median is therefore not merely to fix a value that shall be representative of a set but also to establish a dividing line separating the higher from the lower values. It has one great advantage over the arithmetic mean where the upper and lower values of a set cannot be exactly defined; as in the following data relating to family sizes in a community centre:

Family size	*Number of families*
less than 3 children	6
3 "	4
4 "	3
5 "	2
over 5 "	2

There are 17 families and the median size will therefore be the same as the size of the eight ranking value. The eighth family is included in the second category and the median family size is therefore 3 children. This fact can be stated even though we have no information as to whether families in the first category had 1 or 2 or even no children or, indeed, how many children there were in the two families in the last category.

A median can therefore be fixed even where, as in the above case, the ends of the data sets are 'open', whereas the calculation of the arithmetic mean is subject to some mathematical defects. To calculate the mean it is first necessary to add the individual values together. How does one add x and y together if x is 'less than 3' and y is 'more than 5' when x can be any positive value lower than 3 and y may have any value greater than 5 (subject to reasonable limits according to the nature of the measurements)? This conundrum may be resolved only by arbitrary decisions which might well destroy the utility of the value that is derived from the calculations. Some difficulties do arise in fixing a median value when the available information details the number of values within classes without quoting the actual values, although it is usually enough to be able to identify the class containing the median value.

Some indication has now been given of the relative merits and defects of the various measures of central tendency. Consideration also needs to be given to the effect of changing circumstances upon statistical means. A mean calculated for one set of circumstances will not necessarily apply in different circumstances; nor may it bear any direct proportional relationship to the new mean unless every individual value of the new set bears the same proportional relationship to its respective value in the old set. A motor lorry, which was restricted to a maximum speed of 20 m.p.h., in fact achieved an overall average speed of 16 m.p.h. When the maximum permitted is increased to 30 m.p.h., does the actual average speed increase in the same proportion from 16 to 24 m.p.h.?

To assume that a direct proportional change would result is to ignore the reasons why it is impossible for the lorry to achieve a steady average equal to the maximum permitted. The main reasons are involved in traffic congestion and crawling speeds over varying stretches of road. If it is assumed that the lorry is to travel a distance of 100 miles, then at 20 m.p.h. it would cover this distance in 5 hours, but at the actual speed of 16 m.p.h. achieved it takes $6\frac{1}{4}$ hours. The lost time is therefore $1\frac{1}{4}$ hours. If all this loss is due to traffic congestion then it is possible, though not certain, that if the lorry travels over the same route it may still be subject to the same delays in congestion, even though it could achieve much greater speeds between congested areas if permitted a greater maximum speed.

The lost time may therefore remain the same, being related either to mileage travelled or route taken (both of which are unchanged) rather than to speeds. The new average resulting from the increased maximum speed may therefore be calculated:

100 miles at 30 m.p.h. =	$3\frac{1}{3}$ hours
plus lost time	$1\frac{1}{4}$,,
Total time	$4\frac{7}{12}$,,

whence the average is equivalent to 21·8 m.p.h.

But all the lost time may not be due to congestion; some of it will be lost as a result of reduced speeds for other reasons and a reduction in speed will have a greater 'loss' effect on the faster-moving vehicle. Consequently the true average will be lower than 21·8 m.p.h. These assumptions can only be estimates and they exclude certain other possibilities. If two vehicles set out at the same time and from the same departure point, the faster vehicle will reach another point before the slower one. At different speeds they will reach specified points at different times, so that one might reach a normally congested area before the traffic has become sufficiently dense to cause congestion.

The faster vehicle might therefore experience less lost time or it might merely encounter the congestion further along the road. This, incidentally, is very much what happened in the scheme to stagger working hours in London. Statistics were produced to show that a proportion of workers in one area should finish work a quarter of an hour earlier than had previously been the practice. As a result they certainly avoided the 5.30 p.m. travel crush within their own area only to find themselves caught up in the vortex of whirling humanity elsewhere. Since it can take half an hour to cross London during the rush period and since the latter period is an extended one, the changing of working hours by as little as a quarter of an hour does little to alleviate the travel problem.

The only satisfactory way to calculate the new average speed of lorries under changed conditions is to ignore the original data and collect new details applicable to the new conditions. Any other course can provide nothing but conjecture.

Care must also be exercised when average performances are being compared for ranking purposes. The ranking of clubs in the English cricket championship is decided by the average number of points gained from all its matches to date. Owing to the difficulty of arranging matches and the fact that not all teams play the same total number of matches in the year,[1] it often occurs that at a particular date no two teams may have completed the same number of fixtures. During the season, therefore, the position might be as follows:

	Number of fixtures	*Total points*	*Average*
Team A	7	76	10·86
,, B	9	96	10·67

Team A is ranked first because its average is higher. Both teams gain 2 points in their next respective matches – what effect does this have on the averages? At first glance, although both averages will clearly be reduced, one might not expect this development to

1. Some teams play 32 matches; others play only 28.

have a material effect on the respective rankings. But it does! After the additional matches the table reads:

	Fixtures	*Points*	*Average*
Team B	10	98	9·80
„ A	8	78	9·75

Thus, although their achievements in the additional matches were identical, Team B now has the higher average and has ousted Team A from the top position. Again, one might have expected that, as the two points gained by A would have to be 'spread' over fewer fixtures than those gained by B, the amount per fixture (i.e. the average) resulting from the two points would be relatively higher for A and that, therefore, A would increase its lead fractionally. But, as is shown above, this certainly does not happen.

There are two points of interest here. Firstly, the final result is not peculiar at all; it only appears to be odd because we intuitively expect something else and it is this expectation which is at fault. There is no true comparison between teams A and B at all, since they have not yet played the same number of fixtures. Team B has played two more fixtures and has therefore had greater opportunities for scoring – or not scoring – whereas Team A's opportunities are to come, and no one can forecast their outcome.

The second point is that intuition is proved wrong because it assumed that the addition of two points would increase each average, whereas it actually reduces them. It is therefore not a gain but a loss which has to be spread and this proves more serious for Team A because it has fewer fixtures over which to absorb the loss. The fundamental difference, however, is still the difference between the numbers of fixtures completed. The respective averages cannot be expected to behave in the same manner until parity of matches played is established. To talk of teams changing positions in the championship throughout the season is meaningless if they have not completed the same number of matches; to do so is to compare two measures which are not strictly comparable.

Each form of central tendency measure gives a different conception of average. Which is the best to use in a particular case – the mean, the mode, or the median? The answer, generally, is none of them but all of them. Any one measure alone can give only one dimension of a set of data. To use it alone is like looking through a keyhole; the part of the room you can see cannot give a full idea of the whole room. There is no overall 'best' form of average; so much depends upon which type of variable is being considered and upon the range of the values. Where the distribution is normal,[1] the mean, mode, and median will all coincide but not all distributions are normal. One measure may appear to give a better representation for one purpose but not for another, and it soon becomes apparent that, since these measures are really concerned with different dimensions, it will give a better indication of their true significance if we use them in close conjunction with each other.

It is also necessary to know something about how the values in a set behave, both within their full range and also in relation to the central measure. This is achieved by calculating the standard deviation from the mean, and is discussed in Chapter 15.

1. See Chapter 15.

6 The Persuasive Percentage

Absolute statistics by themselves rarely have very great meaning, but they may take on a new significance when compared with other data or within themselves. A company's sales statistics may produce a rising trend line on a chart and so, perhaps, give the impression that the company is flourishing. But what if, at the same time, other data show that the profit is decreasing? This gives quite a different picture but, if it is accepted that only like things may be compared, how can sales and profits be compared since these are not similar? Well, they are or should be connected and that is a help. For instance, they may both be measured in money terms and so can be compared in that medium. Secondly, the one should be a function of the other; actual profit should increase as the sales increase and in normal circumstances should increase at a greater proportionate rate.

But it is sometimes desirable to compare sets of data concerning two or more variables which cannot be expressed in terms of a common unit. For instance the number of articles sold is measured in terms of quantities, while the profits arising from their sale is expressed in terms of money. Again, it may be desired to compare two variables between which there is no known correlation, the object of the comparison perhaps being to attempt to establish the existence of such a correlation. In order that such a comparison may be effected it is first necessary to transform the data sets into terms of a common dimension. If this is not possible as a direct transformation, a convention is often adopted of relating values within the sets to each other as percentages of the total of the set and then comparing the resultant percentages between sets.

I have called this a convention rather than an actual transformation, because the percentages, although having an appearance of similarity, are nevertheless dependent upon the basic character of a set of data. They are therefore percentages of different things. The percentages within each set are calculated quite independently of the other set; the calculations are, so to speak, turned in upon themselves and this inbreeding may lead to some very odd offspring.

Everyone likes a percentage. To many people the % sign holds the same kind of final significance as the mathematical tailpiece 'Q.E.D.'. It is a symbol signifying an answer to a calculation which someone else has worked out for them. An answer is better than a problem any time. All one has to do is to glance at a column of percentages representing separate quantities and at once one can place them in their respective order of magnitude and therefore according to their relative importance. This may be true of percentages within a set but it does not necessarily follow as between sets. The percentage sign has a comfortably persuasive air of respectability and finality, but it lends itself to uses where its respectability becomes open to doubt.

The following comparative figures show the numbers of cars carried respectively by British Railways Channel boats and by a charter air line in the two years 1949 and 1958, expressed to the nearest thousand of cars:

	1949	1958	1958 *as* % *of* 1949
Railway	73·8	197·9	268
Airline	2·6	67·5	2,596

These figures show that in 1958 the number of cars carried by air was nearly 26 times as great as those carried by air in 1949 whereas, over the same period of time, British Railways had hardly managed to treble their figures. The percentages alone would give a spurious superiority to the airline's share of the traffic. The rate of increase is indeed phenomenal. The airline has made excellent progress; it has, in fact, gone a long way, but the actual figures show that it had a long way to go. 1949 was its first full commercial year, while British Railways had been in the business for many years. To compare the growth of the airline's business with that of the railways is like comparing the growth of a baby with that of a youth.

The percentages do not show that the airline captured any of the railway's clients, but they do show a remarkable increase in the total number of cars being taken abroad. Nevertheless the figures show that, despite the airline's growth, the railway boats still carried three times as many cars as did the airline and that, during the ten years, the railways attracted 124 thousand new clients whereas the airline attracted only 65 thousand. On the other hand it is possible that both services were operating at full stretch at peak periods and that the actual figures for 1958 reveal the respective carrying capacities of the two services rather than their comparative popularity among travellers. The main usefulness of these statistics to either service is not necessarily to compare their relative achievements so much as to keep close watch on each other's activities and also to watch for trends in changing demand for their services. British Railways, for instance, used this information to support an attempt to buy a financial interest in the airline!

Percentages often tend to hide significant aspects of the original data. Local authorities in depressed areas will be concerned about any increase in the absolute numbers of unemployed in their area. It will be no consolation to them that their unemployed figure, expressed as a percentage of their total number of inhabitants, is lower than that of other areas or is lower than that originally accepted as a target figure. Percentages do not provide food. Every increase in the absolute numbers is of direct concern to local authorities. But to a central authority which has to decide which area most needs assistance then, so far as this is measurable at all, the percentage basis is likely to have a strong influence on the decision. It must not be overlooked, however, that a large area with a relatively low unemployment percentage may in fact have more men unemployed than the number of men out of work in a small area having a higher percentage.

Comparative percentages have a much more reliable significance if they represent proportions of the same total quantity. Thus, if the component items of cost of production of a commodity are expressed as percentages of the total cost, any changes in the percentages will at once reveal the need for investigation.

When dealing with such changes, however, it must be realized that a percentage change in one item of cost may alter all the other individual percentages. If there is a simple cost structure in which components A and B each account for 50% of the total cost, then the reduction of A to 49% will automatically increase B to 51%. The actual percentage change in each item is only 1% but the changes are in opposite directions so that there is now a 2% difference between A and B. This can happen even though the actual cost of B does not alter at all:

Component	*Cost* (£)	%	*Altered cost* (£)	*New* %
A	10	50	9·6	49
B	10	50	10	51
Totals	20	100	19·6	100

The cost of B has remained the same but the reduction in the cost of A has altered B's percentage. This means that one cannot accept the new individual percentages as being evidence of anything other than that there has been a change. To discover the nature of the change, it is necessary to go back to the actual costs. Percentages should therefore be accepted only as spotlights focused on particular aspects of data and not as substitutes for the data themselves. Equally important is the rule that percentages prepared for one purpose should not be used for another purpose without ensuring that they can hold the meaning ascribed to them. Figure 14 is a tabulation of the December sales achievements against budgets for a number of different departments within one company. Figure 15 gives similar details for January in the following year.

In December the percentage achievement for the firm as a whole was 98% with some departments showing achievements in excess of 100%. In January the statistics reveal a sharp fall in achievement to 73%. When the company's executives, who have been accustomed to seeing monthly statements on the percentage levels revealed in December, suddenly observe the low achievements in January, their possible reaction is one of panic. Ruin stares them in the face. Should they start a new advertising campaign or employ more sales representatives?

The first commonsense thing to do is to look at the actual data instead of the percentages. This at once reveals that sales have

	Department	*Budget* (£000)	*Actual* (£000)	*% Achieved* (*approx.*)
DECEMBER	A	95	97	102
	B	187	176	94
	C	76	81	107
	D	280	268	96
	E	17	20	118
	TOTAL	655	642	98

Fig. 14

	Department	*Budget* (£000)	*Actual* (£000)	*% Achieved* (*approx.*)
JANUARY	A	149	100	67
	B	250	185	74
	C	100	82	82
	D	389	279	72
	E	30	21	70
	TOTAL	918	667	73

Fig. 15

actually increased slightly in every department. The significant fact revealed by the percentages therefore is not that sales have fallen off but that they have not increased as much as was expected. This paradox again arises because the two sets of percentages are calculated on different bases. They are calculated on the respective budgets and these are quite different, the January one being about 40% higher than the December one. Here is the basic cause for the misunderstanding. The compared percentages have been accepted as illustrating a trend in the level of sales, when their sole object is to measure achievement against the budget and thereby to measure the efficiency of the budgeting process.

It is not suggested that the company is free from worry. The budget will have been used as a basis for other calculations throughout the company and a great deal of expenditure will have been incurred in anticipation of the budget being achieved. Higher rates of production, introduced to match the expected sales level, may have produced a greatly inflated stock figure at the end of January, and the prospect of continuous unplanned stockpiling is a serious one for any company. But the right answer to the company's problems will not be found by the wrong interpretation of data. In the present case perhaps the company ought to look for a new sales manager – not because the sales are falling but because his forecasting may be irresponsible.

On the other hand, it may not be a fault attributable to the sales department at all. It may be that the production department had failed to make the goods which were to have been sold in January. A look at production data as well as at figures for orders outstanding and orders received will make it easier to put the problem in its right perspective. The position may not be nearly so disastrous as first appearances might tend to show. A check on the budgets for other months may show that these are lower than the January budget; a forecast that these budgets might be exceeded would help to reassure the management that the achievement for the year as a whole will more closely approach the budgeted figure. This can happen particularly where orders are received at irregular intervals and for large amounts; all that may have happened is that an order expected in January was not received until February.

Where then, it may be asked, is the value of calculating these percentages at all, if satisfactory answers can always be found to awkward questions. An obvious reply to this is that the answers will not always be satisfactory, but the real value of calculating the percentages is in prompting the questions in the first place. Regular statistics provide a constant probing of the state of the company's business. A headache may be the symptom of one of many diseases or may merely be a passing indisposition of no serious consequence at all. The doctor cannot assess its true significance without looking for other symptoms and trying to trace the cause. In much the same way the percentages are symptoms pointing to one or more possible causes affecting the health of the business. They may be invaluable provided it is remembered that one symptom does not make a disease.

Percentages calculated on different bases are always causing confusion or perhaps merely hiding reality. Some examples are simple to understand; others are not so simple, but they all derive from accepting percentages at their face value without looking at the underlying data. Interest rates on loans are advertised as percentages of the amount loaned, but these declared percentages will not always mean what the reader thinks they mean. One example of this concerns nominal rates of interest payable on loans in circumstances such that interest is calculated on the full amount of the original loan until it is completely paid off, even when repayments are made by equal monthly instalments.

A man borrows £120 at 5% per annum and repays it at the rate of £10 at the end of each month, but he pays interest as if he had the use of the full loan for a whole year. The actual time he has had the use of the money, however, is calculated as follows. When he repays the first £10, he has then had the use of that amount for 1 month. Similarly he has the next £10 for 2 months. His total use reckoned in sums of £10 is therefore $1 + 2 + 3 \ldots + 11 + 12 = 78$ months. But £10 borrowed for 78 months is equivalent to borrowing £65 for 12 months. In effect, therefore, he is paying £6 interest for this latter equivalent and this represents a real annual rate of interest of about 9·23% or nearly twice the advertised rate.

It may not always be clear that the base for percentage calculations has changed. Wages are cut by 10% in one year and increased by 10% in the following year. Therefore the workers are receiving their original rate of pay? Not at all. A person earning £20 per week will receive £18 after a 10% reduction. If he then receives a 10% increase he will receive £19 16s. 0d. and he is still four shillings short of his original pay. In order to restore him to his original pay, his new wage of £18 would have to be increased by one-tenth – approximately 11%. The reduction is 10% of £20. The increase is 10% of £18.

A similar set of circumstances arises where surcharges and discounts are shown on the same invoice

A		*B*	
Price	100	Price	100
less 10%	10	less 10%	10
	90		90
plus 10%	9	less 10%	9
	99		81

Two different calculations are given above. In column A is a plus and minus percentage combination which, while adding and deducting nominally equivalent percentage rates, restores only 99% of the original price. In column B is a combination of two deductions each of 10%, which however together make an actual total deduction of only 19% from the price.

Other sources of confusion are perhaps not quite so obvious, even though they should be. A manufacturer supplied goods to the invoice value of £1,000 less 5% discount, giving a nett value of £950. A further credit note was raised for £60 so that the nett amount to be paid was £890. The total discount was therefore £110 or 11% of the original value. It was later discovered that an error had been made in that the invoice had originally been calculated on out-of-date prices. Prices had in fact been uplifted by 5% and, as this had been omitted from the invoice, the customer derived its full benefit. The total discount shown on the invoice and credit note was 11%. Then the total effective discount is 11 + 5 = 16%. Is this correct? It is not. The original amount of £1,000 should have been uplifted by 5% to £1,050. The amount actually paid was £890 so that the effective discount is £160 and is equivalent to approximately $15\frac{1}{4}$% of £1,050. This is $\frac{3}{4}$% lower than the incorrect value shown above.

This may seem a small margin of error but small percentages have a habit of affecting other relevant figures with quite different results. A manufacturer, whose expenses are fixed in the short term, experiences a fall in sales turnover. His sales in the first year were £10,000 and he makes a profit of £225, the equivalent of $2\frac{1}{4}$% on turnover. The next year his expenses remain the same at £9,775 but his sales drop $\frac{3}{4}$% to £9,925. His profit therefore falls to £150. The position then is

	Sales	*Profit*
Year 1	£10,000	£225
,, 2	9,925	150
% decrease on Year 1	$\frac{3}{4}$%	$33\frac{1}{3}$%

A reduction of $\frac{3}{4}$% in sales turnover is associated with $33\frac{1}{3}$% reduction in profit.

In the above example the fall in profit was expressed as a percentage of Year 1. Why not of Year 2 instead? This is not a simple matter to explain, but a further example may be useful. The value of a variable at one reading is 39 and at a later reading it is 78, so that the difference between the readings is 39. Has the value increased by 50% (i.e. of 78) or by 100% (i.e. of 39)? So much depends upon one's point of view. Sometimes it depends upon what one expects but it primarily depends upon what the percentage is measuring. In terms of actual change, since we are using the first reading as a yardstick with which to compare the second reading, it is logical that we should express the difference in terms of that yardstick also. The value has therefore increased by 100%.

But there could be some complication if one is measuring actual readings against expected values. In this instance it depends entirely on what one expects, as is shown below.

Case 1	*First reading*	*Second reading*	
Expected value	78	78	
Actual ,,	39	78	
Achievement	50%	100%	(Difference = 50%)

Case 2			
Expected value	39	39	
Actual ,,	39	78	
Achievement	100%	200%	(Difference = 100%)

The problem of the shifting base is very much involved in statistical indices,[1] while percentages on different bases are always being called in as evidence in labour problems. The following are the profits and wages (in £ thousands) paid by a company over five years:

Year	*Profit*	*Wages*
1	170	850
2	220	900
3	270	950
4	320	1,000
5	420	1,100

These figures could be represented by an ordinary line chart as in figure 16. Every increase in profit has been matched by an increase in wages, and the lines are always parallel. An employer's argument might be that this is fair to both sides and that, if the chart shows anything else of importance, it is that much more is paid in wages than is paid in profits. The workers have nothing to complain about.

The employees would justifiably take a very different view. The data represented in figure 16 take no account of the payment per individual. Suppose that there are 1,000 employees and 200

1. See Chapter 12.

shareholders; then, in Year 1, the average pay per employee was £850 and each shareholder 'earned' the same amount by way of profit. If the numbers of employees and shareholders remain the same, then in Year 5 the average pay per employee will have risen to £1,100 but the average profit earned per shareholder will have jumped to £2,100. Wages have risen by about 29% whereas profits have increased by about 147%. The employees would therefore

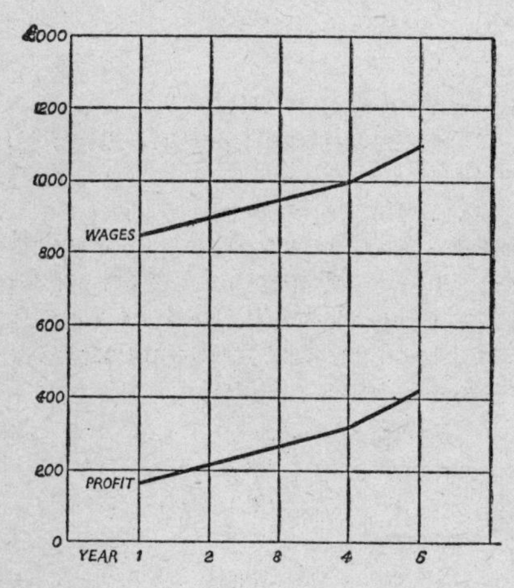

Fig. 16

prefer to see a chart (as in figure 17) showing the percentage changes, since this will emphasize that profits have leaped away ahead of wages.

Both charts record the truth; which is one to accept? This selection will usually be affected by bias. If you are a shareholder you will undoubtedly prefer figure 16. It may be that profits were too low in Year 1 and that they have just reached a reasonable level. But you may, in any case, feel that your class is entitled to the same absolute increase as that received by the employee class since profit is payment for risk-taking; if the firm had made a loss then it is your money that would have been lost, whereas the employees would still have received their wages. You might, however, pause to consider whether the relative payments would be equitable. The same principle would be involved if, in times of famine, each family ought to have two loaves of bread irrespective of the size of the family. In a family of two, each person would have one loaf; in a family of eight, each would have only a quarter of a loaf. The employee cannot be blamed for saying that this is not the same thing; a family of eight is not the same as a family of

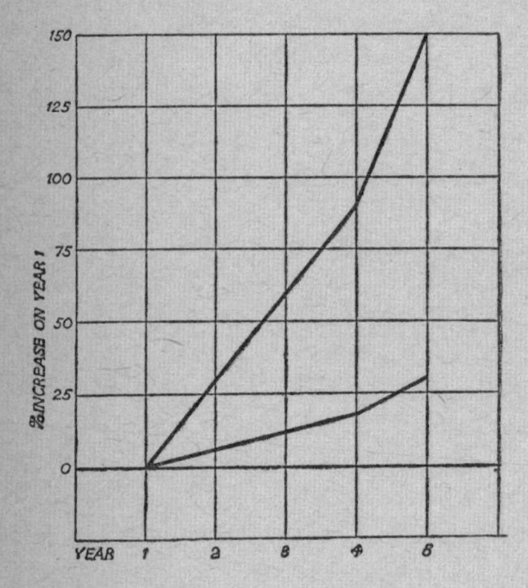

Fig. 17

two. He will therefore accept figure 17 which first shows what has happened within each class before relating the results between classes.

The statistician, however, will accept both charts. They do not contradict each other. They merely reflect different aspects of the same things in the same way as a camera may photograph one aspect of a concert hall while a microphone may record another and quite different aspect. Space science has now made it possible to photograph the other side of the moon; the results are quite different from the photographs to which we are accustomed but it is the same moon. The charts serve to demonstrate what has happened. They are not intended to prove what *should* have happened. The fact that they may be employed to illustrate different points of view does not guarantee the correctness of the views themselves.

Percentages also figure in wage disputes between different types of labour. Unions for skilled workers are concerned to keep their members' wages at a level higher than that of unskilled workers. This differential may be expressed either in money terms absolutely or as a percentage on the lower wage. When the time comes for all-round increases, how are these to be calculated? If the skilled worker at present receives ten shillings more than an unskilled worker and they both receive a further ten shillings increase, then the skilled worker will still have maintained his differential. But, if the skilled man is content with a ten shillings differential when the respective rates per week are £10 for a skilled worker and £9 10s. for an unskilled one, will he still be happy in the years ahead when the rates have increased to £15 and £14 10s. respectively? The ten shillings advantage will then represent a less significant part of his income and its value in real terms will have lessened so that the differential, although nominally the same, will actually have narrowed. The skilled man will therefore say that overall increases should be expressed as percentages of basic pay so that his differential also increases. This is reasonable, but would he also agree that the same principle should be applied to everyone? A 10% increase on his wages of £10 will give him another £1 per week, but the same increase rate applied to the salary of a managing director will, if the latter's salary is £10,000 per annum, bring him in another £1,000 or nearly £20 per week.

The uses to which percentages may be applied obviously have their complications. There are also restrictions upon the mathematical treatment of percentages. A percentage is a derived statistic just as is the arithmetic mean, although it is derived differently. Means cannot normally be added together and then averaged to produce the mean for a number of data sets, and there are similar restrictions applicable to percentages. The latter may be treated together arithmetically only if they have been derived as calculations upon a common base or if the individual percentages are identical with each other. The following figures show expected values and actual values of a variable; and the latter are then expressed as percentages of the former.

Variable	*Expected*	*Actual*	%
X	50	30	60
Y	70	35	50
Z	80	60	75
Totals	200	125	62·5

It is not possible to derive the percentage relationship between the totals by adding the individual percentages and dividing by 3. This would give an incorrect value of $61\frac{2}{3}$%. The true value may be calculated only from the totals themselves.

The individual percentages are calculated to different bases and therefore have different identities as if they belonged to different families. They cannot therefore be legitimately added together at all, quite apart from what can or cannot be done once they have been supposedly added. Yet it is not uncommon to see reports of price increases which 'add up' in this way. Manufacturing costs may be quoted as having risen as follows: 'manufacturing expenses by 3%, trading expenses by 6%, wages and salaries by 5%, materials by 1%; therefore total costs have risen by 3 + 6 + 5 + 1 = 15%.' This, of course, is nowhere near the truth. What the correct answer really is we cannot say without knowledge of the actual costs, but we do know that the highest percentage increase is 6% and that the overall increase cannot exceed that figure. There is also another possible fallacy which may have been built into the statement – the fallacy of double-inclusion. Wages and salaries, for example, may already have been included either as manufacturing or trading expenses and, if this is so, then the statement has in effect included them twice for good measure.

Percentages may be expressed in different ways so as to give different degrees of emphasis. If a second value is expressed as a percentage of a first value, the growth which has occurred between the two values will appear to have more significance than a percentage which expresses the actual growth. If the values of A and B are 189 and 267 respectively then B is equivalent to

141% of A, whence the growth is equivalent to 41%. If B is shown as having grown to 141%, the impact on the mind is greater than if it is shown as having grown by 41%.

This is a very subtle influence, but it may take a number of different forms. Dividends on share capital are usually declared as percentages of the latter. Thus a 5% dividend will produce £5 income for every £100 capital held. If the dividend is increased to 10%, this transaction may be represented in more than one quite distinct way. One method is to report that the dividend has increased by 'five percentage points' – that is from 5 to 10%. Thus a capital of £100 at the beginning of the year will have grown to £110 (including dividend) instead of to only £105. But the impact may be magnified by reporting the change in the dividend as having been doubled.

If a shareholder's dividend receipts jump from £3,000 to £6,000 per annum, it would be reasonable to present this as an increase of 100%; the impact of this representation would be justified by the size of the values involved. If, however, his dividends increased from £1 to £2, it would still be true that his income had doubled, but the difference in actual money would be so slight that to express it as a 100% increase is to accord the change a significance which it does not really hold. The shareholder has little cause for rejoicing.

It will be seen that familiarity with the % symbol does not necessarily imply an understanding of its uses and meaning. Usually it is the differing base which, appropriately enough, is at the bottom of much of the trouble. Differences in the levels of percentages themselves may appear to be minimized or exaggerated merely by the method of treatment accorded to them. Percentages may often help to bring out the significance of a set of data and, indeed, may sometimes be too revealing for one's comfort. On the other hand, they may serve to encourage misplaced emphasis and should be accepted only in association with other statistics. A lone percentage is as uninteresting as an unaccompanied soprano; the accompanist is as essential to the recital as is the soloist herself.

7 A Sense of Proportion

Percentages are representations of proportions and ratios, but there are other ways of expressing these without recourse to the % symbol. The calculation of proportions is generally effected in order that, once calculated, they may be used in further applications. This, as has been mentioned in the previous chapter, may be dangerous unless one also has regard for the original data. This is of paramount importance in statistics.

Mathematically the proportions $\frac{9}{10}$ and $\frac{90}{100}$ are equivalents; the second reduces to the first by cancelling out the noughts. Another way of stating these proportions is as 'nine out of ten', it being normally implied, for any set of data containing more than ten values, that this proportion is really an average proportion for the whole set. Any statement which contains a phrase of this nature should be carefully scrutinized for its credentials. It may, of course, be phrased differently as 'nine times as many . . .' and, although it is often purposely phrased in this down-to-earth manner in order to bring home truths to the public without the use of percentage symbols, its use is often statistically indefensible.

The first question that should be asked is what is the size of the population which is supposed to have the characteristic claimed for it. It will often be found that this sort of statement is based on hunches and has no statistical support at all. If the population is large or difficult to contact, it may safely be assumed that the statistic is derived from a sample since it is unlikely that all the members of the population will have been contacted. At once we know that the statistic is subject to sampling errors which should be considered.[1] Having found the size of the population, what is

1. See Chapter 16.

the size of the sample? Can the statistic derived from the latter properly be applied to the population? This same question arises even if the nature of the population is not explicitly stated since the existence of the population is implied.

One often sees reports, particularly in advertisements, that 'nine out of ten' men prefer a particular proprietary article. These reports may be the results of properly conducted surveys, or they may not. The cynic will ask who the ten men were, and unfortunately his scepticism may well be justified. Nine doctors out of ten said to favour a particular treatment gives an impression that nine out of *every* ten doctors had the same preference. The original statement omits the all-important word *every* but it does not refer to the fact that perhaps only ten doctors were asked. Most people, however, assume that such a statement refers to the total population of all doctors and it is unwise or dishonest to issue such statements unless they do so refer to a population.

To reduce proportions to equivalents expressed as tenths or hundredths may be misleading in other ways. If a pupil is awarded 18 marks out of 20 in an examination, where one mark is given for each correct answer, then he has answered two questions wrongly. But 2 out of 20 is the same as 1 out of 10 when reduced to its lowest terms. If, however, in a different examination under the same conditions, he is awarded 9 marks out of 10, he has again forfeited one mark but this time he has only one wrong answer. He has achieved the same proportion of marks in each examination but has answered more questions incorrectly in one than in the other. The reduction of a proportion to its lowest terms thus restricts the interpretation of the final expression. The twenty-question examination results should be expressed either in full as '18 marks out of 20' or as '90%' since the latter makes it clear that the expression is a proportion.

While it is legitimate to use the percentage expression in this way, it is not necessarily permissible to extend this to the expression of proportions as, for example, '94 out of 100'. This is mathematically the same as 94% but in its particular form of expression it may appear to mean something else. If a pupil's marks are 47 out of 50 possible then, proportionately, this is the same as 94 out of 100. But since it was impossible to score 100 it is incorrect to express the results as being part of that figure. It would suggest that the pupil had dropped six marks instead of only three. The apparent total possible is in excess of the actual total possible. A proportion should be expressed as 'out of' another figure only when that figure actually represents the total population, unless it is made clear that the proportion, as reduced, is an average one for the population.

Even where this requirement is fulfilled, however, the results may sometimes be misleading. In a report on the hat-wearing habits of the British male, the *Financial Times*[1] gave a headline 'Only one new hat in eight years', which, to a casual reader, suggests that every man buys a hat every eighth year: that is, that the replacement rate of hat buying is once in eight years. But it does not mean that at all, as the more careful reader will find if he reads the full article. This stated that the sales average out at only one new hat every eight years for men and youths over sixteen. The figures cannot be used to provide a replacement rate because nobody knows how many of the total population of men and youths do not wear hats. We know how many hats are sold and an estimate of the present population size, but we do not know how many men actually buy hats and we therefore cannot deduce how often they buy new ones. This does not lessen the value which the underlying data have for the hatters; it is merely the expression of that data which is at fault.

This kind of loose phraseology is often encountered. Whereas 'one in every ten men' might suggest that out of every ten men one and one only would possess the characteristic claimed, so 'every fifth man' conjures up a vision of men being selected at regular intervals. But characteristics will certainly not be found to repeat themselves with unfailing regularity. Members of populations cannot be filed away with such precise tidiness. Similarly the assertion that 'every minute a man is knocked over in London' is also basically at fault. Leaving aside the comedian's comment that if a man is knocked over every minute, then it is time he got out of the way, the point is that even though there may be sixty accidents in every hour it does not follow that there is one

1. 10 December 1959.

accident every minute. There may be two accidents in one minute and no accident in the next.

All that any of these statements means is that there were a certain number of occurrences affecting a certain-sized population or a certain period of time. Within one year, for example, there are so many minutes of time and also so many recorded accidents. It is the relationship between these two sets of data which the statements seek to express. That they do not do this successfully is entirely a matter of defective phraseology involved in the attempted compression of ideas which, perhaps, are not compressible. A paragraph which gives a précis of a book cannot be expected to give the whole story. Similarly, some statements cannot be compressed to the extent which is sometimes attempted. Statisticians avoid such expressions whenever possible. While it is true that the expressions often arise out of an honest attempt to put the figures over to the public – although the attempts are not always honest – it is precisely because they are intended for the public that the statements should be clear and not subject to the misinterpretations that have been noted.

If an ordinary individual reads that there are 525,000 accidents in a year, the figure alone may not mean very much to him. It is a worthwhile endeavour to point out that this is almost the same number as there are minutes in a year. So that, measured against time, there is on average one accident for every minute. But this merely reveals a one-to-one correspondence between the individual members of two populations. The relationship is purely numerical and helps to provide a mental fix of the size of the populations. It does not mean that for each pair (i.e. one minute and one accident) the individual occurrences are simultaneous or related in any other way.

Other popular statements of extremely dubious worth are those typified by the statement that, if all the banknotes in the world were placed in one stack, the stack would reach as far as the moon. They wouldn't, of course, because the stack would collapse before a few thousand notes had been stacked. What such a statement tells anybody is anyone's guess. It gives no idea of the number of notes or of the value represented, but merely tells us that the figure is very large. Presumably, if there had not been enough notes to facilitate this comparison, they might have been stretched end to end or folded across the middle. There is no stopping the space artists. The futility of their comparisons was once parodied by a wit who said that if all his friends at a bottle-party were stretched end to end on the floor he wouldn't be at all surprised! The parody was probably much more closely related to reality. The human mind generally cannot distinguish between two astronomically great numbers. If readers are told that the stack of all banknotes reached as far as the moon and that the pile of all coins would reach out to Venus, they would know that both involved very large quantities but few would be able to say whether the stack was therefore higher than the pile or vice versa.

It is possible to differentiate between large numbers which are not too great, but an interpretation of the difference also requires a sense of proportion. Sheer weight of numbers does not necessarily prove anything, although many issues of the day are decided by majorities. If the majority wish to take a certain action this works quite well in practice, but it may well be that, in certain circumstances, the minority's views are ignored to the detriment not only of themselves but, in the long run, of the members of the majority also.

Numbers for the sake of numbers alone is most undesirable in statistics. In the amusing case of the Twelve Red-Bearded Dwarfs[1] heard before Mr Justice Cocklecarrot, counsel for the defence proposed to call seven thousand witnesses, whereupon Mr Snapdriver for the prosecution threatened to call more than twelve thousand. Anything the defence could do, the prosecution could do better! There was no suggestion that any of the 'witnesses' knew anything about the court case.

A sense of proportion is equally essential when dealing with data of the 'more' and 'less' categories. Bed is not the most dangerous place merely because more people die in bed than elsewhere. In the nature of things it is because they are dying – or at least because they are ill – that people take to bed in the first instance. Again, there are more people in London than there are in Bath who cannot read, but this does not mean that Londoners

1. J. B. Morton, *Diet of Thistles*, Cape.

are generally less educated than the inhabitants of Bath; it merely reflects the fact that there are many more people in London anyway.

Much the same fallacy occurs in attempts to compare the number of intoxicated pedestrians involved in accidents with the number of intoxicated drivers similarly involved. There are many more injuries to intoxicated pedestrians than there are to intoxicated drivers. So it is safer to be a driver! If you become inebriated, all you have to do is to borrow somebody's car and you thereby increase your chances of survival! The population of drunken pedestrians of course greatly exceeds the population of drunken drivers and that is why there are more of them who sustain injuries.

These examples are perhaps too obvious. Others may not be so obvious. A newspaper report[1] gave a representation of accident data related to speed which, the motoring correspondent asserted, proved that speed limits did not cure road accident incidence since it was safer to drive at higher speeds. The data are as in figure 18.

This shows that the lowest involvement rate was attributed to vehicles travelling at between 55 and 70 m.p.h. and highest under 40 m.p.h. The involvement rate shown is the number of vehicles involved in accidents per 100 million vehicle miles of travel. It is too easy to look at the graph and jump to the conclusion that it is safer to drive at 80 m.p.h. because there are less accidents at that speed than there are at 35 m.p.h. But a closer look should make it clear that this interpretation is invalid. The graph shows accident data and not data relating to the 'safety' of different speed levels; these are two quite different things. If it be contended that it is 'safer' at 80 m.p.h. than at 35 m.p.h., then, by the same method of comparison, it is also 'less safe' at 80 m.p.h. than at 60 m.p.h., at which point on the graph the curve commences to rise again. The contention that higher or lower speeds *in general* are safer is thus self-contradicted.

There is a further sound reason why the graph alone cannot be reliably interpreted without reference to supporting data. No

1. *Evening Standard*, 10 June 1960.

involvement rate at all is shown for vehicles travelling at speeds below 35 m.p.h. Speeds in this category therefore cannot be compared with higher speeds and consequently the whole concept of meaningful comparison is destroyed. Again we are seeing only a part of the picture.

The statement that there are relatively fewer vehicles involved

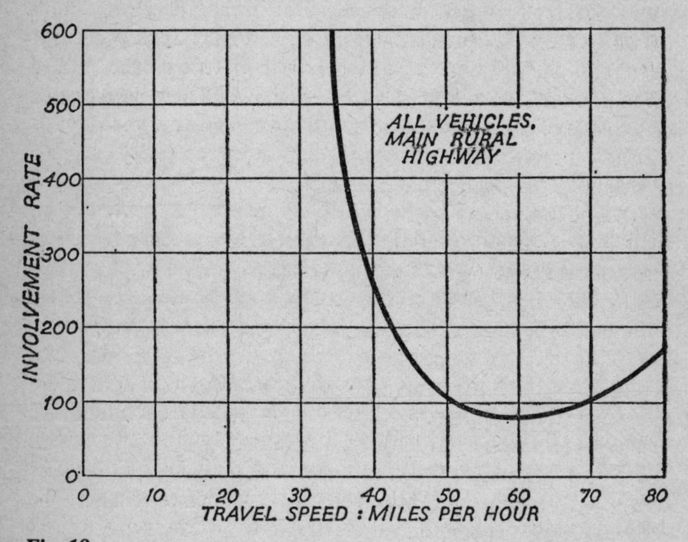

Fig. 18

Accident involvement rate of vehicles (reproduced from Traffic Engineering, *Washington D.C.)*

in accidents at 80 m.p.h. than there are at 35 m.p.h. is not synonymous with the contention that 80 m.p.h. is a safer speed. It is almost certain that the involvement rate for vehicles travelling at 80 m.p.h. is low merely because the number of miles travelled at that speed is relatively low and that the number of accidents is correspondingly low because of the lower level of operation at this speed. There are fewer vehicles involved in accidents at this speed partly because there are fewer vehicles actually travelling at this speed. A colleague, in discussing this set of data, commented that perhaps there were less accidents at the higher speeds simply because in order to achieve such speeds a car must first work its way through the lower speeds; there were so many accidents at the lower speeds that the vehicles could not reach the higher speeds and therefore could not be involved in accidents at the higher speeds! There is also a further interesting point to consider; presumably not all of the 80 m.p.h. vehicles were involved in accidents with other vehicles travelling at the same speed. Thus one of the 80 m.p.h. cars may have been involved in an accident with a 35 m.p.h. car and, indeed, the accident may not have been caused as a result of the operation of the cars at the stated speed levels at all. In compiling data to assess the 'safety' of different speed levels, one should include only those accidents which can reasonably be attributed to speed. If the wheel of a car travelling at 40 m.p.h. works loose and thereby causes that car to collide with another car travelling at 70 m.p.h. it would obviously not be valid to attribute the collision to the speed of the second car.

Similarly, data which show a decline in the number of offenders convicted of crime do not necessarily justify the contention that the crime wave is receding. The data may possibly reflect improvements in police preventive methods, so that the number of crimes had in fact declined, but without more detailed information they could just as easily reflect a much less satisfactory state of affairs – less convictions may have been made simply because the police had failed to apprehend the criminals. In order to interpret the data it is necessary to have a great deal more detail – how many reported crimes were there? – what percentage of these resulted in arrests? – these and other facts must all be taken into consideration. Why is the incidence of teen-age crime, as a percentage of all crime, increasing? Is it because the nation has entered a moral decline, or is it merely that the teenagers now represent a greater proportion of the total population?

The importance of background information cannot be too strongly emphasized. A report on road accidents in 1959 showed that 'motor cyclists and scooter-riders suffered the greatest increase in casualties during the year'. In 1958 the number of fatalities in this category was 1,421, whereas there were 1,680 fatalities in the following year. This was an increase of 259 deaths, equivalent to 18·2% of the 1958 figures. The inference which might be drawn from this was that this form of transport was becoming much more dangerous but, looked at more penetratingly, the statistics tend to behave themselves.

	1958	1959	*Increase*	% *increase*
Number of vehicles (in thousands)	1,520	1,733	213	14·0
Number of fatalities	1,421	1,680	259	18·2
Fatalities as % of vehicles	0·093%	0·096%		

Although the number of fatalities increased by 18·2% much of this is attributable to the fact that the number of vehicles in use increased by 14%. The nett change in fatality data expressed as a percentage of vehicles on the road was only 0·003%.

There are many examples of this tendency to accept the mere size of the numbers involved as having special significance. In 1960 the British Medical Association carried out a survey of the incidence of illness among railway clerks. The investigation covered the record of 1,260 Scottish clerks between the ages of 45 and 64 years and it was found that the following average absences were recorded:

Chief clerks:	Between	2 and 4 days
Senior ,,	,,	7 and 12 ,,
Ordinary ,,	,,	11 and 15 ,,

A newspaper reporter read into these figures the fact that promotion appeared to help in defeating illness because the clerks in the higher grades took less time off from work. This might encourage the lower grades to try harder for promotion but it is to be feared that they would be sadly disappointed. The more responsible individuals will take less time off from work and the more responsible individuals are also, it is be hoped, those most likely to be promoted. Another possibility, of course, is that once having achieved promotion, the more senior members of the staff could not afford to be 'ill' too often – there might be too many others waiting eagerly for their jobs. A chief clerk might well feel constrained to return to duty before he was fully recovered, whereas an employee with less responsibility might not. The absence figures cannot prove anything as to the degree of severity of illness suffered.

It is a popular pastime to bring up large numbers as artillery to blast through opposition to arguments. Every time there is a major strike in an industry, someone starts to calculate what the strike has cost. Obviously it will have cost something in terms of lost profit, but the additions are often apt to be a little odd.

Arguments in favour of the adoption of a new world calendar have been supported by a 'well-known radio commentator's estimate that our present system of reckoning time costs the City of New York approximately $5,322,886·25 annually, devouring the taxpayers' dollar at the rate of $607 per hour'.[1] How did he manage to calculate this? He reckoned that, with the present calendar, it would take about one minute to find out the day of the week on which a specified date would fall. He then assumed that half of the population of New York would carry out this calculation once a day, giving a total time lost of 58,333 hours daily. At 25 cents an hour, this total is equivalent to a loss of $14,583·25, and the final figure is reached by multiplying by 365 to give the total loss for a year.

There are three assumptions upon which this most original calculation is based – that it takes one minute to think of a date; that everybody does it once in two days; and that a financial loss is incurred as a result of the thinking time. The first two assumptions are obviously guesses. How could these facts possibly be known? But whatever the merits of these two assumptions, the third is nonsense. How can cost be brought into the calculation – cost of what? Perhaps it is a figure for the wages which the State of New York's employees have obtained by false pretences since, instead of boosting production, they have taken time off to ponder on the all-important question of whether the 4th July fell on a Sunday last year. They may, however, have done their pondering during their lunch break or, even if they did it during working hours, it may have been done at the expense,

1. E. Achelis, *Of Time and the Calendar*, New York, Hermitage.

not of production, but of a discussion on yesterday's baseball game.

In view of the sweeping assumptions incorporated in this particular exercise, the precision of the final answer adds the final touch of absurdity that cannot but give the statistician cause for hearty laughter – particularly the quarter dollar – and it's a poor heart that cannot rejoice occasionally!

A similar misunderstanding arises in commerce. A problem may be posed as follows. The average cost of preparing, typing, and despatching an invoice is approximately three shillings – is it then worthwhile to render an invoice to the value of two shillings and elevenpence, since the invoice will cost more than the amount one hopes to recover. A little thought will show that this last assumption is incorrect. The average cost of an invoice is calculated by adding together the cost of stationery, salary of invoice typist, postage, establishment charges, etc. and dividing the total by the total number of invoices sent out. The rendition of one extra invoice, however, will not cost another three shillings. The establishment charges and the typist's salary are already paid. The cost of one additional invoice therefore consists only of the postage and perhaps threepence for stationery.

This sort of calculation falls into the same category as those humorous ones involved in proving that nobody works at all. The year has 365 days but, as a worker works only eight hours a day, each day is worth only one-third of a day in terms of work performed. The worker therefore works for only 122 days; deduct all the week-ends and holidays and the worker is left owing his employer a number of days.

In statistics, or any other science for that matter, one cannot, merely because they are there, subject numbers to mathematical processes just as momentary whims or axe-grinding demands. A woman, who decided not to buy a coat at a sale, proudly informed her husband that she had thus saved £15 – the purchase price of the coat. She could not, however, appreciate his suggestion that she might lend him £5 and keep the other £10 for herself; nor did she take too kindly his subsequent claim that, although she had saved only £15, he had saved £600 by not buying a car. Juggling with numbers does not necessarily constitute mathematics, any more than does the 'proof' that, if one apple keeps the doctor away, then two apples should keep two doctors away.

Mathematical processes may only be used if they are relevant to the problem. A glimpse of the obvious? It is remarkable how often such glimpses pass unobserved. Everyone alive today had two parents and each of these had two parents, so that everyone now living had four grandparents. Does this mean that there were four times as many people living in their day as there are now? It does not, of course; it would only be necessary to go back a few more generations to show that an individual had more ancestors then than there were people alive in the whole world. The point here is that ancestors of different individuals do not necessarily form mutually exclusive sets. Children of parents who are themselves first cousins, will have only six great-grandparents – not eight, since they will have a common descent from two great-grandparents. Very large numbers often contain an element of exaggeration, if not of definite inaccuracy.

So much for large numbers. A sense of proportion however is just as essential when dealing with small numbers. Twice nothing is still nothing. Compare the following profit figure for two departments of one firm – in £ thousands.

	1959	1960	*% increase*
A	30	36	20
B	1	2	100

Department A has increased its profit by £6 thousand and yet shows only a 20% increase whereas Department B has doubled its profit. But the Department B profit has increased only by £1 thousand, and it would be ridiculous to point to this department's percentage rate of increase as a target for Department A. Again, a small firm, which supplies only one customer, will increase its number of customers by 100% when it finds its second customer. The presentation of the proportional figures without details of the absolute statistics would obviously be misleading. For another example, one might read that 25% of a firm's employees under twenty years of age had been dismissed in one week. This might point to a major labour problem until one realized that, in fact, only one man had been dismissed – there were only four employees under twenty years of age.

The use of proportions may also be misleading even if the figures themselves are not. They may give a reasonable idea of the proportional make-up of a population, but how are they to be used? Often enough their assumed significance is shaded by the interpreter's own preconceived ideas and expectations. If he considers a 30% proportion important he may perhaps say: 'there are *at least* 30%...' whereas, if he considers it unimportant, he may say: 'there are *only* 30%...'

A proportion of one population may indeed be more significant than the equivalent proportion of another population; or the proportion levels of one population may have different meanings for different sections of the community. A condemned man may say that he has only two days to live whereas a child, who has to wait two days before some exciting event, will feel that he has an eternity to survive. These are extreme cases but there are many intermediate ones.

This shifting emphasis needs to be watched most carefully. The question as to how the statistic may be used will be a question of fact in each particular case. A manufacturer discovers that 30% of his total sales are effected in one town whereas, of other towns, not one accounts for more than 2%. It will be important to him to know that 30% are accounted for by one town, but he cannot ignore the fact that 70% of his sales are accounted for elsewhere. The truth is that he probably cannot take any action at all on these figures alone, but they serve to give one reflection of a particular circumstance which may be fitted into place in a panoramic representation of a whole set of circumstances. Only thus may the proportion perfectly fit the perspective.

8 What's in a Name?

The object of statistics is to measure facts and probabilities so as to draw out their inner meanings, but before one can say what a set of data *means*, it is first necessary to be certain that one knows what it *says*. The claim that figures speak for themselves is often quite unjustified; it may well be necessary to translate them into a language understandable to others or so to rearrange them that they speak more loudly. But, again, in order to translate from one language to another it is essential that the translator thoroughly understands the meaning of words used in the original language.

This is also a basic requirement in statistics. The variables measured must be clearly and precisely defined. Obviously one cannot measure anything unless one clearly understands its limits, and those who have not made the measurements must, before they utilize the derived measurements, ensure that they also have a precise image of the thing measured.

'"When *I* use a word," Humpty Dumpty said in rather a scornful tone, "it means just what I choose it to mean – neither more nor less." '[1] Before Humpty came to his notoriously sticky demise, he had successfully placed his curse upon humanity – for it is a real curse upon the spreading of modern knowledge that so many misconceptions may arise from the different meanings or shades of meaning of words. Terminology has become so confused that one often has to ask the question, 'What do you mean by that?' The English language abounds in words with multiple meanings so that a man may appear to say one thing and yet mean something else. On the other hand, many different words possess much the same meanings and this is made even more confused by

1. L. Carroll, *Through the Looking Glass.*

the fact that Americans and Englishmen, although nominally speaking the same language, often describe objects by different names. Yet there are enough problems without these national differences, as is typified by the example of the pair of English verbs 'to best' and 'to worst', both of which mean 'to get the better of'.

In statistics it is not merely, or even always, that main word meanings become confused so much as that there can be so many shades of meaning within a word group. It is essential that the qualification of each term used is known and thoroughly understood. What, for example, does one mean by the cost of living? Few people who use this term would be able to give a clear indication of its real meaning, yet probably everyone uses it or carries some idea of it. Is the cost of living really the cost of buying the essentials wherewith to sustain life or does it include luxuries? Before deciding this point one must first define a luxury. In Great Britain a few years ago a refrigerator was clearly considered a luxury but it is not similarly regarded today. Does the cost of living include every item of expenditure so that taxation enters into it as well as direct food purchases? Is it really a concept of cost at all or is this 'cost' really based on the somewhat different aspect of price levels?

In Great Britain people often refer to the cost-of-living index having moved up or down a number of points – yet there is no such index! The index referred to is the Index of Retail Prices.[1] By its nature it does, of course, reflect movements in nominal cost levels but the reflection is an indirect one since it derives from an averaged expenditure list for all households. This at once gives a qualification to the word 'Prices' and there are many other qualifications to the meaning of the index – so much so that a special booklet has been issued to explain them.[2] This explanation constitutes the definition of the index and the latter is meaningless without it. The index is, for example, made up of a number of main sections each of which has its own index number and these

1. See Chapter 12.

2. *Method of Construction and Calculation of the Index of Retail Prices*, H.M.S.O.

are related to each other so as to calculate an index number for the complete set. When reference is made to the index, therefore, it is possible for confusion to be caused by the fact that the reference could be to the index number of a section instead of to the main index number.

Similar difficulties arise in other countries as well, and when one wishes to compare the cost of living (having once established what one means by that) in different countries, it will be found that their governments have quite different ideas as to its meaning. Indices in different countries are based on different methods and different interpretations. Before they can be compared they must first be transformed into terms common to both – and this may well be impossible.

Another example of international differences is provided by statistics referring to the number of British holiday visitors to European countries. If the figures supplied by the separate countries were added together, the total for 1959 would be at least 4½ million whereas the British Travel and Holidays Association estimated that the total was approximately 2¼ million.[1] This difference arises because of the different definitions of 'visitor' used in the separate countries. Italy, for instance, counts her visitors at the frontiers and so includes tourists who are passing through to another country where they will be counted again. Switzerland, however, counts the hotel registration, thereby omitting visitors, such as campers, who do not register at an hotel, and including more than once those visitors who stay at more than one hotel during their holiday. Other countries have other methods again.

Perhaps the cost of living is too complex a concept to define at all. But confusion can exist in everyday matters and often with the most simple words. What is a factory? If one is given a figure said to be the total of factories in Britain, what would this mean? A company has two buildings in one estate, is this one factory or two? Two adjacent companies merge their business and open up doorways in the existing walls to allow movement between their premises. The two premises in effect become one – does this

1. *The Economist*, 25 June 1960.

convert two factories into one? In fact, a company may have different buildings for different production processes, but for many statistical purposes each separate building is counted as a factory. Information such as this is a prerequisite to the interpretation of data purporting to record the number of factories and their various characteristics.

Figures showing the production rate of domestic ladders are useful only if one knows what is meant by the qualification 'domestic'. Ladders are classified in this way in Great Britain for the purpose of purchase-tax levies and are not necessarily used for domestic purposes at all. They are classified as domestic merely because they are of a size more likely to be used for domestic purposes. It was once pointed out that a 12 ft ladder was cheaper to buy than was an 11 ft ladder, simply because the latter was classified as domestic and was thus subject to purchase tax whereas the larger ladder was not!

In the last two examples the differences in meaning are quite intentional. They are different because they are meant to be. They have been made different either to assist in the compilation of data or in order to fix limits for taxation or other purposes. These qualified definitions must be understood and may usually be found with regard to official data if one takes the trouble. The Central Statistical Office issues a *Monthly Digest of Statistics* which contains a mass of data on many different subjects. The necessity for clarity of definition will become apparent when it is known that in addition the Office issues another booklet[1] containing seventy pages, the purpose of which is to explain the meanings of the various terms used and the variables represented in the data.

Official terms are often intentionally restricted in meaning while words and expressions in everyday use also acquire different shades of meaning by virtue of the use made of them by manufacturers and others. This is particularly noticeable with regard to size. A colleague once asked in a restaurant for 'a little chicken' hoping for a few slices off the breast. Instead he was given a whole chicken; it *was* a little chicken but it was not what had been

1. *Monthly Digest of Statistics Supplement – Definitions and Explanatory Notes*, H.M.S.O.

expected. It was a good thing, he thought, that he had not ordered 'a little turkey'.

At the other extreme, words like *giant*, *family*, *household*, or *handy* as being allegedly descriptive of sizes of product packs have lost their ordinary meanings and, indeed, now have no measurement meaning at all. A giant packet of one detergent powder may be the same size as a large packet of another. Packet sizes for different products therefore cannot be treated as equivalent in volume merely because they are expressed in the same terms, particularly when it is noted that a 'full' packet is not always full. A full packet is one in which the manufacturer has packed the standard amount of contents, but it is often not full in the real meaning that the packet is filled to capacity. The size of a packet is no guarantee of the 'size' of the contents.

If comparison is being made between the relative economies of different detergents, one should compare the actual amounts of detergent in each packet and then check for relative washing efficiencies. The packings should not be compared at all, yet a large packet printed in certain colours will attract buyers merely by its appearance. Statistical packings are the apparent meanings of terms used; the contents are the actual meanings employed for those terms in each separate instance.

It is the contents which count. A good example of this is given by the results of a survey made by the Consumer Advisory Council[1] of the prices of tins of peaches. To most shoppers there are perhaps only two sizes of tin – large and small – but it is not sufficient to compare the prices of a number of 'large' tins and to say that the most expensive is the least economical. Some results of the survey are shown in figure 19.

It will be seen that there were only slight differences in the gross weights of the unopened tins. The prices per tin varied much more widely, although to a shopper the tins would all have appeared to be much the same. A very different picture, however, is drawn with the relative weights of fruit content – that is, the real commodity purchased without its packing. The highest content measurement was shown by tin A (65%), whereas tin H

1. *Shopper's Guide*, February 1960.

had only 48%. Tin H, which was cheaper per tin than tin A, was nevertheless more expensive in cost per pound of fruit. The cheaper tin was more expensive!

These differences do not, of course, take into consideration the quality of the fruit. Weight of fruit is not necessarily a guide to value, but how does one define quality in such a way that it may

Tins	*A*	*B*	*C*	*D*	*E*	*F*	*G*	*H*	*J*	*K*
Total contents (oz.)	17	$16\frac{1}{4}$	$16\frac{3}{16}$	$16\frac{7}{8}$	$16\frac{9}{16}$	$16\frac{9}{16}$	$16\frac{11}{16}$	$16\frac{11}{16}$	$15\frac{3}{4}$	$15\frac{5}{8}$
Fruit contents (oz.)	$11\frac{1}{16}$	$10\frac{7}{16}$	$10\frac{3}{16}$	$10\frac{3}{16}$	$9\frac{15}{16}$	$9\frac{1}{16}$	$8\frac{3}{8}$	$8\frac{1}{16}$	8	$8\frac{1}{2}$
= % of total	65	64	61	60	60	54	50	48	51	54
Price per tin (pence)	30	27	24	25	$17\frac{1}{2}$	26	26	26	23	28
Price per oz. of fruit (pence)	2·7	2·6	2·4	2·4	1·8	2·9	3·1	3·2	2·9	3·3

Fig. 19

be measured? Is it the condition of the fruit, its colour, its texture or food value or merely its taste which makes it a better quality purchase than another; and whichever of these is the real criterion, how is it to be measured? It just cannot be measured scientifically. A personal preference for a particular commodity may be very real but it is not possible to measure exactly just how great is that preference over other similar commodities. The preference, however, may be real and, although it cannot be measured, it can be counted and the totals of preferences of a group of people may provide very useful information provided the limitation of definition is remembered.

The comparison of tins of peaches also illustrates another aspect of definition. Not only must the variable itself be identified, but the unit of measurement which is applied to it must also be defined. Variables must be measured in appropriate terms. Area, for instance, cannot be measured in inches. The most practical measuring unit for tins of peaches is the cost of fruit per unit of weight, not the overall cost per tin.

Definitions having been clearly fixed and understood, it is always necessary to ensure that definitions do not change either during the process of data collection or between the processes of collection and application. Alternatively, if the definition remains identical in changed conditions, consideration should be given to the question of whether the definition is still a realistic one in the new circumstances. This is the main weakness of every statistical index which is based upon the fiction of a non-changing society.[1]

Definitions themselves are often misinterpreted in the actual collection of data and, since each interpretation is in effect a separate definition, such misinterpretations will provide sets of data based on differing definitions. This is particularly true where observations are recorded in a number of centres for, if categorization is difficult, it is hardly to be expected that a number of people, making their records quite independently of each other, would reach the same decisions with regard to possibly equivalent circumstances. A striking example of the possibilities of such situations is provided by the widely diverging sentences imposed by different magistrates for approximately similar offences. It may be said that some of this divergence is due to differing circumstances in each separate case brought before the courts and that it is not entirely due to magisterial inconsistencies. This, however, cannot alter the fact that there are differences.

Variables which are being measured for statistical purposes may also be affected by differing circumstances but, once they have been recorded and incorporated in the general data, the variables in effect comport themselves as if they were all homogeneous. Whether they have been affected by differing circumstances or other factors is not primarily of importance if their dissimilarity is clear. The fact of this dissimilarity is important since dissimilar objects cannot be added together to provide any total other than one which represents a collection of dissimilar objects. No amount of mathematical ingenuity can help in adding x and y to give any

1. See Chapter 12.

total more simplified than $x + y$, unless they can both be expressed in terms of a third quantity.

Different hospitals recording cases of a certain disease may be asked to classify these into four categories – critical, serious, average, or minor. Each of these classifications is subject to misinterpretation despite the very precise definitions which may be provided in support. No two medical cases are ever exactly alike and a decision has to be made in respect of each separate case. If the person who judges the seriousness of the case is also the person under whose care the treatment will be effected, then it is possible that his judgement of a case will be affected by his own ability as a doctor. The rated seriousness of the case will depend ultimately upon the chances of a cure and this, in turn, depends partly upon what the doctor thinks he can achieve in that particular case. The more confident a doctor is of success so the less serious he may consider the case. Different doctors may therefore be employing standards which also differ and the figures they submit will not then be comparable. This difficulty will, of course, be likely to have greater effect on cases which lie at the margins between classes.

In the tabulation:

Critical cases	28
Serious ,,	55
Average ,,	101
Minor ,,	210
Total	394

the only reliable figure may be the total, although there is also the possibility that misinterpretations might tend to cancel each other out. It may be that the final result is not so very different from the true state of affairs after all, but some definite information is required on this point before the figures may be used with any confidence. This kind of information is obtained by instituting some form of checking procedure into the machinery of a statistical survey. Thus, where two persons are engaged in recording measurements of a nebulous type of variable, their duties might be switched occasionally to ascertain whether any differences in recorded measurements which they may make can be attributed to bias.

In some cases definitions must of necessity be couched in general rather than specific terms, but this will inevitably weaken the general applicability of the results achieved. This is quite noticeable in the present practice of staff-grading for salary purposes. It is perhaps possible to adopt a dozen classes of employee, from general managers down to office boys, and to classify each position in a company within these broad classifications, taking into account the responsibilities of the position-holder. It is, however, questionable whether one should classify the position held rather than classify the employee who fills that position.

Individuals are not so easily classifiable as are their nominal responsibilities largely because their nominal and actual responsibilities are not identical. No two jobs of the same description are ever exactly alike and different individuals will make those jobs even more different over a period of time either with or without the assistance of Parkinson's Law. In a large firm a cost accountant's functions will probably entail much more serious responsibilities than will those of a cost accountant in a smaller firm. In a still smaller firm the cost accountant may also have to double as general office manager or shoulder other administrative responsibilities. Yet they are all cost accountants and therefore nominally in the same classification. Staff grading procedure usually allows for grading within classes according to responsibilities, but how could the cost accountants in each of these three examples be satisfactorily distinguished from each other in order that they be remunerated according to their work output?

Differences of opinion on this subject are at once apparent. In 1960 the Office Management Association published a scale of grading which included cost accounting and its concomitant responsibilities in the highest clerical grade. The editor of an accountancy journal,[1] however, very properly commented that these duties should be carried out by a qualified accountant and not by unqualified clerks at all.

1. *Cost Accountant*, April 1960.

Suitable classification might be possible if it were to be effected by one person who could weigh up the individual responsibilities. But this is never so in fact, since different firms carry out their own grading. Even within firms it is normal practice for departmental chiefs to grade their own staff. Bias and prejudices may result in unbalance as between departments, and it is a solemn thought that some of the people who carry out the grading have no idea of the difference between the medians and the means of salary distributions published by their obliging trade or employers' association. Furthermore, many employees enjoy certain hidden advantages such as expense allowances, use of a car, and other perquisites in addition to their nominal salaries which are included in the data. These extras all come within a definition of income and to omit them from comparative statistics is unrealistic. This is a related problem to that of assessing social classes by salary or other income levels. Today this is a most haphazard procedure as was exemplified by a newspaper correspondent who pointed out that, as a catering officer, he received much the same salary as did a young typist. Some manual workers earn as much or more than some salaried staff.

The author is aware of one instance where all the members of the staff of a company were graded according to a procedure suggested by an employers' association, and it was thereupon discovered that many members were underpaid in comparison with the reported salary levels of other companies. Here, one might have thought, was a reasonable case for an all-round increase, provided one accepted the schedules published by the association. Not at all! The grading officer thought that he must have made a mistake and he promptly regraded the whole staff! This selectiveness of definition represents abuse of statistics at its worst. If you dislike the answer, just alter the definitions. If it even occurred to the officer concerned that in regrading the staff he was also degrading himself, perhaps he was able to soothe his conscience with the thought that the published statistics didn't mean anything anyway. But had they suited his purpose he would presumably have produced the published data as evidence.

There remains the problem of the shifting definition. Data

gathered over a period of time, although nominally representing different measurements of the same variable, may in fact at times represent measurements of different variables. Statistical data are useful only if the nature of the variable measured does not itself alter. If an inspector is measuring and recording dimensions of steel bolts which are coming to him on a conveyor and if, suddenly by some production freak, the conveyor starts to bring him brass nuts, he would not measure the latter in the same way so as to combine the measurements of the steel bolts with those of the brass nuts. A brass nut can never be a steel bolt. But the changes which occur on the production line of statistical data are usually much more subtle and certainly less remarkable. You can recognize the difference between a steel bolt and a brass nut. You cannot, however, always clearly distinguish between two statistical variables which are apparently similar, neither can you always know whether there is a difference at all.

Reference has already been made to the shifting-definition problems in indices in general. Difficulties arise because the structure of the relationship between the variables measured tends in time to be no longer representative of the concept it allegedly represents. The same difficulty, however, can arise also with the more elementary price relatives,[1] where there is only one apparent variable to be measured. If the nature of the variable changes, then subsequent measurements cannot be directly related to measurements made prior to the change. A manufacturer concentrates upon the production of electric cookers and has only one design of model in production at any given time. When, after twenty years of successful operations, he looks back over his past he decides that he would like to calculate what retail price increases have occurred on his products. What, in other words, is the price of an electric cooker today compared with the price of an electric cooker twenty years ago?

The first question which the statistician must ask him is what does he mean by retail price changes and then, no less important, what does he mean by an electric cooker? The manufacturer will probably reply somewhat icily that surely he knows what an

1. See page 147.

electric cooker is, and that retail prices are what the customers have to pay for the cookers. It is as simple as that. Only a statistician would quibble, but it is just as well that he does.

The model of cooker which was on sale twenty years ago is not on sale now; likewise, the model which is now on sale was not on sale twenty years ago. No one can measure the retail price of a non-existent cooker and it is therefore impossible to trace the changes in one individual model over the full twenty years. The design of domestic appliances is changing annually and in twenty years there may have been five or more different models succeeding each other on the production line. They would all share the characteristic that they were electric cookers but might not bear any real resemblance to each other in appearance, efficiency, or the indefinable quality variable. The first cooker produced may have sold for £10 and the present model may retail for £40. Then it would be true to say that cookers were now four times as expensive as they were, in the sense that it would cost a purchaser £40 today instead of £10. But apart from this very limited meaning, it is obviously incorrect to compare these two figures to gain an idea of increases in retail prices measured, for instance, per unit value of utility purchased by the customer.

The only way in which this might be attempted would be to measure total price changes within the production life of each model and to relate these changes to each other by means of conversion factors.[1] This will give an approximation to the desired answer, but it must be borne in mind that, even within the production life of one model, a number of modifications might have been introduced so that the model, although nominally the same throughout, may have changed considerably by the time its production ceased.

Most definition problems exist before a statistical survey is commenced, although they may not affect an individual who has to interpret them until a later date. The problem should therefore be tackled before the survey is begun. Most important of all is that the object of the survey should be clearly defined. It is all part of one process. The ideal state of affairs is one in which there is

1. See page 164.

no lack of precision; where everyone expresses himself clearly, says what he means, means what he says, and, if there are any doubts, expresses them. This is a very unusual state indeed! When asked to provide and interpret data the statistician must protect himself by asking as many questions as are necessary to identify the real problem. Many inquirers are not entirely clear as to what they want to prove but hopefully believe that 'the figures will show *something*'. There is never any guarantee that the statistician will be able to produce an answer for the inquirer – the latter may be asking for the impossible – but his task is made that much more simple if he understands the question!

If he does not protect himself by first clarifying the true requirements of an investigation, his report is liable to be returned with a note to the effect that he has not given certain details – and it is too late then, however satisfying, to send a note back saying, 'You didn't ask for them.' The statistician should always be in at the start of the race and, even though he may not be able to make his own terms he should ensure that everyone understands the terms that are used. Otherwise the race is just a handicap with no prizes for anyone.

9 General and Approximate

Whatever particular form of statistic one selects as being representative of a set of data, this is but the first step. The next step involves the question of how to use it. The first important requirement in this connexion is a full realization that the statistic used represents (if anything at all) the data from which it is derived and nothing else. By the correct application of statistical method this actual representative of a sample may be converted to an estimated probable representative of the population from which the sample was drawn, but it will have changed its identity in the process.

Provided the calculations have been effected correctly the mean of a known distribution is something definite and unalterable since the data from which the mean is derived are themselves fixed. They are measurements which have been made and cannot therefore be altered. But when this same mean value is accorded the status of an estimate of the population parameter it loses its exactness. It is impracticable to record all the measurements of every member of a large population – that is why a sample is taken in the first place – and, since we do not have all these measurements, it is obviously impossible to calculate an exact representative value for them. It is not that there is no exact answer, for theoretically there must be one if the population is finite; it is merely that we cannot calculate it. But with the assistance of probability theory the statistician has been able to measure the probability that the actual value of a sample mean will within limits correspond to the value of the population parameter. It is scarcely to be expected that these will correspond exactly and the correspondence is necessarily approximate but, since the probabilities involved may be measured, it is possible to take a number of steps to ensure that possible errors are minimized. The approximation may sometimes be brought so close that its element of error is of reduced importance.

Statistics is built upon approximations. Measurements which can easily be accomplished throughout a whole population provide material only for descriptive statistics. The real art of statistical inference is called into action only when exact data are not available, such as statistics on a national basis which are not available merely because no organization exists for their collection. In such circumstances, statistical statements can only be approximations. There is no weakness in this. If you ask someone for an approximate answer, you cannot condemn him because his answer is an approximation. One of the troubles in statistics, of course, is that very often one person will ask the questions for a specific purpose, but someone quite different will try to use the answers without reference to that purpose and without asking how nearly exact those answers are.

There is therefore very rarely an exact result to be derived from a statistical investigation and this is one very great difference between the processes employed by the statistician and by the accountant respectively. They both have to do with numbers and both have to interpret the implications of the numbers produced. But, whereas the conventional accountant keeps track of every penny which passes through the accounts, the statistician will round off the figures to the nearest hundred or nearest thousand pounds. Very often the accountant's figures are the raw materials for the statistician's work; the latter is able to round off the figures with confidence only because he knows that the accountant has meticulously guaranteed their accuracy.

Approximations are more common in this life than is generally supposed except by those who have to do the measuring. There are many more shades than just black and white. All scientific measurements are to some degree inaccurate, either because no measuring device can record an exact actual reading or because of the experimenter's inability to read thc index accurately. The length of a line for instance can never be measured exactly even though it may normally be expressed as an exact number of inches. If the width of the marking line itself on the ruler is 0·02 inches, it follows that what we read off as 1 inch may in fact be anything between 0·99 and 1·01 inches. This is a possible error arising from the construction of the ruler.

There is also a possible observational error in the visual reading of a length. This will vary as between individuals and it is possible that in some instances it may cancel out the effect of the constructional error but it is more likely to aggravate it. The closer the limits of error are together so is the measurement that much more accurately representative of the true value which, though difficult to measure, nevertheless exists. For many purposes, however, it will not make any practical difference if what we call 1 inch is in reality 1·01 inches. The odd hundredth of an inch may not make any real difference and, to all intents and purposes, what we read as an inch is taken to be an inch in actual fact.

Even exact mathematical results may be only approximate in relation to real life since they are based on the assumption of a continuity of the circumstances assumed to exist for the purpose of the calculation. If the circumstances alter then so may the true value alter. It is worth noting here that very often the population studied may itself be changing in content and that if it is very large it will be impossible to measure it exactly since it will be altering during the actual process of measuring. By the time that national census figures are issued, for example, the demographic population statistics will have altered because of the births and deaths which have occurred in the intervening period. Yet, at a specific time, the population is finite and, if the changes are comparatively small, they will be insufficient to have any significant effect upon the results obtained. As it is realized that the results are to be approximate anyway, it is valid to round off the population figure and to treat this, for mathematical purposes, as if it were truly a finite number.

Many mathematical calculations themselves are limited to dealing in approximations. The area of a circle cannot be calculated exactly since the formula πr^2 involves the quantity π which is itself immeasurable. An English mathematician, William Shanks, has calculated the approximate value of π to 707 decimal places[1]

1. *Proceedings of the Royal Society*, 1873–4. Vol. XXII.

yet, according to another mathematician, Simon Newcomb, only ten decimal figures are sufficient to give the circumference of the earth to a fraction of an inch.

Similarly the value of $\sqrt{2}$ cannot be calculated exactly, yet these quantities may still be used in some calculations which will give mathematically exact answers. The area of a circle is equivalent to πr^2 and the circumference is equivalent to $2\pi r$. Although π is immeasurable, these formulae make it possible for the measurements of the area and the circumference to be expressed as functions of each other. The use of approximate values in a calculation does not lead to an approximate answer if they may be cancelled out. But even when the answer is itself an approximation the derived answer may be so close to the real value that there is no practical difference between them. Practical utility triumphs over unattainable accuracy.

Close approximations are thus very often good enough for our purposes and statistics helps to define the limits within which such approximations function. It is essential that their identity as approximations should always be clearly kept in view. Yet, how often is this particular requirement lacking from the popular interpretation of statistics. Herein lies the main source of misunderstanding. A business executive, who should have known better, upon being told that a sample survey (sample size 4321) showed that 25% of all male inhabitants over the age of twenty in the United Kingdom used an electric shaver, applied this statistic to the total population represented. There are 17,275 thousand in this population[1] and the executive therefore included in his report the statement that 4,318,750 (i.e. exactly 25%) men then used an electric shaver. No more and no less.

This form of spurious accuracy can lead to serious misconceptions as well as to merely ridiculous results.[2] To use an approximate value as if it were an exact one and to employ it in other calculations so as to produce another apparently exact answer is patently absurd. This practice is unfortunately quite common, but it would be much less common if more people

1. As at 30 June 1959. *Monthly Digest of Statistics*, H.M.S.O.

2. See page 97.

possessed a simple basic knowledge of the facts. The executive mentioned above should have known that, for more than one reason, the attainment of exactness was impossible.

A sample statistic will very rarely be exactly equivalent to the true population parameter. A statistician, in presenting his figures, will say either 'that 25% of those interviewed used electric shavers' or 'that the survey showed that there were 95 chances out of 100 that the population proportion lay between the limits $25 \pm (1\frac{1}{3})$ per cent' or words to that effect. The proportion of the population with electric shavers therefore probably lies somewhere between $23\frac{2}{3}$ and $26\frac{1}{3}$%. It is clearly impossible to calculate the exact numbers of shavers in use since it is thus impossible in the first place to be absolutely sure of the population proportion involved.

If the executive had paused to consider his problem properly, he would have noted that the figures for the total population given by the Registrar General are quoted to the nearest thousand and that it excludes youths under the age of twenty. He therefore begins his calculations with an approximate value which understates the actual figure and he should realize that no amount of mathematical manipulation can ever transform an approximation into an exact value. Then he should have queried the statistic 25%. Could it really have been so exact or was it really 24·9% or perhaps 25·1%. The actual size of the sample would have given him a clue. A total of 4,321 individuals were interviewed, of which 25% is equivalent to $1{,}080\frac{1}{4}$. The number of individuals owning electric shavers may have been 1,080 or 1,081; it could not have been $1{,}080\frac{1}{4}$ and was not therefore exactly 25%.

The last two points do not call for any knowledge of statistical method. Common sense is all that is required. When a newspaper reports the size of a crowd as ten thousand, it does not mean that there were exactly ten thousand people present. In the absence of any definite information it would be impossible to distinguish between nine thousand and ten thousand, and this partly explains why different journals will report different crowd sizes for the same crowd. Statements such as these can be identified as containing approximations if one asks how the reporter could possibly know the right number or whether the terms in which the report is written really describe the event covered.

A valuable picture is sold in America and an English newspaper states that 'it was reported to have been sold for £53,571'. This is not correct. Nobody reported that the picture had been sold for £53,571; the original report was that it had been sold for about $150,000, this being an approximation to the actual amount which was later revealed as $150,750. The figure of £53,571 was calculated by someone on the staff of the English newspaper by dividing $150,000 by the approximate conversion rate of 2·80. It is understandable that he should wish to show a sterling equivalent of the amount involved since his readers are naturally more familiar with English money and the news impact will be all the greater. But to convert an approximate figure using an approximate conversion rate to produce an exact figure in sterling is absurd. This particular aspect is not primarily a statistical one but it serves to illustrate the necessity for questioning published figures. Are the original figures exact or approximate? Does the report actually quote these figures or have they been approximated or otherwise converted by somebody else in the meantime?

Just as it is incorrect to accord an approximation the spurious appearance of accuracy, so is it incorrect to select specific figures intuitively from a set of data and to pretend that these are representative. A set of data must in the first instance be treated as an entity and regarded in its proper context even if, at first sight, it appears to contain contradictions. Many seemingly contradictory quotations may be cited from the Bible if they are taken out of their proper context, but such contradictions are in the words only and not in the acts which the words purport to record. They do not affect the fact that, despite them, Christians can nevertheless find a central truth in the work as a whole. Thus to select one particular value merely because it seems to be right – and this usually means that it serves to support a particular argument – and to discard the remainder is dishonest in some circumstances but is always unwise whatever the circumstances.

To generalize from the specific is obviously wrong unless each and every member of a group is identical for the purposes of the survey. Suppose there are three different societies of different sizes and that every man in a town is a member of one and only one society; suppose then that it is known that the members of one society all possess a characteristic which distinguishes them from the members of every other society. If it desired to know how many of the total population possess one of the characteristics all that one need do is to interview one man from each society. One may generalize within each society only because every member is identical to the particular member interviewed. But to say that the average height of men is 5 ft merely because the first two men measured were each 5 ft tall is invalid for the simple reason that all men are not identical. This form of reasoning is not uncommon in less obvious circumstances and many sweeping generalizations are made without any statistical evidence to support them.

It is also wrong to treat a specific member of a population as if he conformed exactly to the measurements of the population parameter. One cannot particularize from the general. The philosophers of ancient Greece were very fond of postulating paradoxes to confuse their rivals. One of these asked how one could reconcile the statement 'that all Cretans are liars' if the statement is made by a Cretan. If they are, in fact, all liars then the person making the statement is also a liar. If he is a liar then his statement is untrue. Therefore all Cretans are *not* liars, but this contradicts the original statement.

Here is a fine example of formal logic getting the better of common sense. It should be clear that if all Cretans were liars then the particular Cretan could never have made the statement attributed to him. He would have been forced to lie and would then have said that Cretans were not liars. The problem posed is unreal. On the other hand, if the statement about all Cretans was really a generalization meaning that 'most Cretans were liars' or that 'Cretans told lies most of the time', then the speaker could have been one of the few who told the truth or he may have been speaking in one of his lucid truthful moments.

Thus, to pose the problem at all the Greeks really had to apply a general characteristic to a particular individual. They should first have made sure that the characteristic was in fact identical throughout all the members of the population and that it was possible for a particular member to take some action attributed to him. They also overlooked the fact that, although it is possible for everyone to be untruthful at some time or other, nobody can tell lies all the time. It is in any case an intriguing thought that in a society where everyone lied about everything then, since everyone would know that everyone else was lying, the lies would unwittingly become inverted though distorted representations of the truth.

An arithmetic mean is calculated so as to provide one dimension of a set of data, but it can do no more. It cannot be accepted as a dimension of any one or more individual members of the set. By studying one sample of individuals and relating them to others of the same kind it is possible to predict reactions of the latter as groups, but it is not possible to put every individual neatly into a box and label it with a complete identity derived from the sample statistic. There is no man-in-the-street, although there are men-in-the-street, a paradoxical case of a plurality without a basic singularity.

A. P. Herbert has drawn attention[1] to the fact that the Common Law of England has been built around a mythical figure – the Reasonable Man – but that, although many legal cases are in the final analysis decided with reference to whether the actions involved were those of a reasonable man, no mention has ever been made in law of a reasonable woman. From this he draws the amusing conclusion that at Common Law the Reasonable Woman does not exist. At the same time, although legally noticed, the Reasonable Man who is reasonable in all things is as mythical as the Reasonable Woman. Every man is at times unreasonable even if only for the sake of enlivening an otherwise dull existence. Similarly there is no statistical or average man.

The average man does not exist. The average applies to a set of data and not to an individual and if the average man could exist he would be such an odd specimen that by his very uniqueness he would deny his own title. An insurance company, when fixing the premiums payable on its policies, is not concerned with its liability to one individual. Instead it is concerned about the

1. *Uncommon Law*, Methuen.

total amount that it may be called upon to pay in benefits to the total number of policy-holders. The average is, however, often misapplied to individuals.

Quite apart from the difficulty of definition inherent in the salary grading of staff[1] it must also be remembered that the salary mean may result from high payments made by enlightened firms balanced by low payments made by others. The jobs are then 'averaged', in the sense that similar ones are classified in the same category and the earnings within each category are then averaged. The result, more often than not, is then related to individual earnings. A company which cooperates in a survey of this nature should use the results to indicate whether, for the various categories, their standard of pay is higher or lower than the levels revealed in the survey.

More frequently, however, the approach is quite different. Mr Brown does get less than the average, it is true, but he still receives more than the minimum, so Mr Brown is not really entitled to an increase after all. If Mr Brown is paid an income equivalent to the average revealed, this is still no reason why he should not be paid more if he is worth more. If a company accepts the average salary as its criterion, it can expect only average performance in the employees' duties. The logical equivalence of averages is dangerously close to mediocrity.

1. See page 109.

10 Cause and Effect

If you depress an electric light switch and the light flashes on, then the depression of the switch is the cause of the effect of flashing. If, however, the front door bell rings and if at that precise second a piece of toast pops out of the electric toaster, did the ringing of the bell, or the electric circuit thus established, precipitate the ejection of the toast? This is not a likely proposition but, if it were considered possible though surprising, the only way to satisfy one's curiosity on the subject would be to try the process out again in the form of an experiment. Place another piece of bread in the toaster and then depress the bell button. If nothing happens you can throw the piece of bread to the birds and not mention the matter to anyone; but if the toaster does react in the same way as before, you may have discovered some freak of nature which you would do well to exploit. Nevertheless you would need to repeat the experiment many times before allowing yourself to be convinced.

The likelihood of such an outcome is so remote that it is doubtful if anyone would hesitate in dismissing the relationship between the bell and the toast as non-existent. Yet it does not always pay to be too sceptical. Equally odd things have happened. An American housewife complained that every time her telephone bell rang it switched off the television set.[1] It was discovered that the pitch of the telephone bell could, in fact, lower the volume of the television receiver which was under remote control. There was therefore a direct causal relationship.

A further example of the non-existence of a causal relationship is provided by the circumstances in which, when the hands of one

1. *Daily Telegraph*, 2 April 1960.

clock point to ten o'clock, the bell of another clock chimes ten times. The mechanism of the first clock does not cause the second clock to strike the hours. Merely because two occurrences are simultaneous, this in itself is not proof that there is any relation of cause and effect between them. The examples quoted are fairly obvious ones, but it is not always so easy to determine whether events are related or not. A new remedy is developed by chemists and, as the sales of the remedy rise, so does the number of cures increase. Are these occurrences related? Is the remedy really a cure?

This proposition not only asks whether the remedy is a cause responsible for an effect; it also asks whether it has a particular effect. The reported incidence of disease may decrease because the symptoms are no longer visible, but the disease may merely be repressed with the promise of breaking out in some other form later. Reported data cannot give proof of a cure. This sort of proof would require repeated experiments under strict controls. There remains, however, the question of whether the remedy has had any effect at all.

In the toaster episode it was suggested that a number of experiments would prove whether there was a causal relationship, and it is probable that only one experiment would be necessary to dispel any doubts on the subject. The clock problem, however, demonstrates that the continued repetition of events alone is certainly not sufficient to establish the existence of a relationship. If both clocks maintain perfect timing and the mechanisms remain fully efficient, then every time the first clock points to the hour so will the bell of the second clock chime.

It is clear, therefore, that to establish a causal relationship it is not sufficient to show that events are simultaneous nor even that they always occur simultaneously. Something more is required. The disease and cure relationship presents more problems than did the toaster or the clocks. Nobody would expect to find a relationship in either of those sets of circumstances but one does, rationally or not, expect a remedy to have an effect on disease. It is a dangerous tendency, however, to accept an apparent relationship just because one expects to find that it does exist.

The number of apparent cures may have increased as the result of many causes other than the use of a specific remedy. The weather may have been more favourable to the building up of resistance to infection. The sales of a remedy for colds may have increased rapidly but upon investigation it might prove that the consumption of lemons had also increased in proportion. The

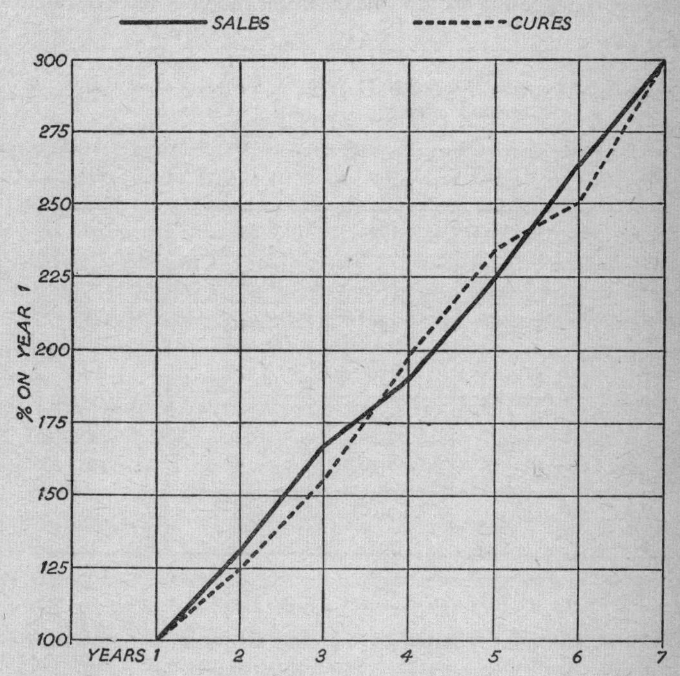

Fig. 20

lemon juice might have been a better remedy than the proprietary article. To which of these or other possible causes is the decrease in the incidence of colds to be ascribed? This again would require a series of controlled experiments but the original apparent relationship serves the purpose of providing pointers to possible cause and effect relationships which may then be subjected to more specific analysis.

There are, of course, a number of ways in which the possible relationship may be disproved, it always being easier to prove a negative proposition. Figure 20 shows the percentage changes in the sales (imaginary) of a specific remedy and the number of reported cures of the ailment for which the remedy is prescribed. The percentage figures are calculated on the data recorded for

Year	*Sales of remedies (thousands of units)*		*Number of cures reported*	
	Number	*% on Year 1*	*Number*	*% on Year 1*
1	30	100	50	100
2	39	130	62·5	125
3	49·5	165	77·5	155
4	57	190	97·5	195
5	67·5	225	117·5	235
6	78	260	125	250
7	90	300	150	300

Fig. 21

Year 1. The values charted are therefore simple index numbers or relatives.[1]

The sales line is almost straight while the cures line, although not similarly regular, fluctuates within very narrow limits about the sales line. At first glance there appears to be good reason for looking for a direct relationship and one turns hopefully to the tabulated data which the chart represents (figure 21).

Each set of data shows approximately equivalent increases year by year calculated on their respective base-year figures. There

1. See Chapter 12.

were fifty thousand cures in the base year as against the sale of thirty thousand remedies which may or may not have contributed their share of the cures. Between Year 1 and Year 7, however, there has been an increase of 60,000 remedy sales whereas there has been an increase of 100,000 cures. There are therefore at least 40,000 additional cures to which the remedy could not have contributed directly and there must be other causes affecting the cures. The increase in the number of cures is not entirely due to the direct influence of the increase in the sales of remedies.

From the data it may be seen that in every year the ratio between remedy sales and reported cures is approximately equivalent at 3 to 5. There has apparently been no increase in the percentage of total cures attributable to the use of the remedy, although it may have maintained its level. The data indicate that the remedy was not responsible for all the increased number of cures. They do not, however, indicate what proportion of cures in any year were due to the use of the remedy nor whether this proportion remained constant from year to year. The so-called remedy could have been entirely useless. The cures may be real but the data contain no proof that any one of the cures stemmed directly from the use of the remedy. Only if it could be proved that there were no other effective causes at work which could have affected the circumstances – that is, if the remedy may be identified in isolation as the one factor – only then would it be legitimate to conclude that the cures were the effect of the remedy. This, however, is not possible, for in every year there are more cures than there are remedies sold and there must, therefore, be at least one other cause. This one other unknown cause may in fact be responsible for all the cures.

Another way in which the significance of the figures may be assessed is to relate them to some other data not shown in the original tabulation. The latter revealed only the number of remedies sold and the number of cures. What about the total incidence of the ailment, as distinct from the number of cures reported; how many cases remain uncured? Figure 22 tabulates the number of cures against some supposed total number of cases. These figures are quite imaginary and are used only to illustrate the points made.

Over the seven years the number of cures trebled, the nett absolute increase being 100,000. The case data now reveal that 50% of all cases were cured in Year 1 but that only 30% were cured in Year 7. Whilst the number of cures had risen by 100,000, the number of cases requiring treatment had risen by 400,000. The number of cases was assuming the rate of an epidemic, rapidly outstripping the rate of cures effected. The absolute

(in thousands)

Year	*Cures*	*Reported cases*	*Not cured*
1	50	100	50
2	62·5	140·5	78
3	77·5	200·5	123
4	97·5	280·5	183
5	117·5	340·5	223
6	125	420	295
7	150	500	350

Fig. 22

statistics for cures showed a remarkable increase over the years, simply because there were many more cases requiring treatment. There has, in fact, been a worsening in total results as is indicated by the lower cure-percentage rate. Although in Year 7 there were 100,000 more cures than in Year 1, there were also 300,000 more uncured cases. The growth in the number of cures achieved does not, by itself, give any indication of the number of uncured cases nor of the impact of the remedy upon the total incidence of the ailment. Different conclusions might have been drawn if the number of cures had risen during a period when the level of cases was falling or was the same from year to year.

The manufacturers of the remedy may point to the figures and say that a reasonable reading of them pointed to the fact that there had been a definite increase in cures and that they could take credit for some part of this increase. Had their remedy not been on the market, then the number of uncured cases might have been even greater. That may be true, but they have not proved their point. The 'remedy' may in fact have had a detrimental effect upon patients and may, indeed, have been the real cause of some of the increase in the number of uncured cases!

But if it is supposedly possible to establish that there was a causal relationship between the sales of remedies and the number of cures, which of these two variables would be the cause and which would be the effect? Distinguishing between cause and effect is not always as simple as in this case. It is often a difficult task and sometimes an impossible one. The bogus relationship must first be excluded. A burglar finds a window unlatched and enters the house, leaving the window open behind him. Is the window open because the burglar is in the house, or is the burglar in the house because the window is open? Strictly speaking, this is not a causal relationship conundrum at all. The burglar may have been in the house because the window was unlatched, for if it had been securely fastened he might not have been able to enter. But the unlatched window was not the cause of his entering; it merely assisted him. Conversely the window is not necessarily open because the burglar is inside, since he may have left by the same means, without closing the window, so that he is not inside at all.

A true causal relationship may have an extremely complex structure. One of the important laws in economics states that prices vary *inter alia* with demand and supply. An increase in the demand for goods, assuming a fixed level of supply, will tend to force prices upwards, but the upward trend will in time tend to reduce the demand since fewer people will be able to afford to buy the goods. The same result occurs when the supply decreases relatively to the demand. The reverse happens when the demand decreases or the supply increases. The difference between demand and supply is either a surplus or a scarcity of goods and each of these has its own effect on prices. Prices in their turn have their own effect on demand and supply. A high price may force consumers to lessen their demand or may encourage manufacturers to increase their supply; low prices may have the reverse effects.

In a limited sphere of activity it is sometimes possible to identify these factors as causes or effects. A manufacturer who reduces his price and thereby increases his sales knows well enough which is the cause and which is the effect. In the national situation, however, except where one supplier or one purchaser (e.g. a wholesale merchant) has a virtual monopoly of the market, it is not true to say that demand or supply or price alone is a causal factor in changes in any of the others. Nor is it true to say that any of these changes are the direct results of changes in the others. Up to a certain point such statements might be true, but a stage is eventually reached when cause becomes effect in the same way as the middle man in a group of workmen handling bricks is a receiver in relation to the man on one side of him but a despatcher in relation to the man on the other side of him. Demand, supply, and price all interact upon each other. The precise limits are difficult to define because the effective mechanism of this interaction is itself not clear and there are many other factors (e.g. the supply of money) which also have an effect and which may themselves be effects in the closely interwoven economic fabric.

The original cause sets off a chain of reaction; it produces an effect which then becomes a cause of subsequent events. Where this development is all in one direction it is easier to differentiate between the various factors, but where the development doubles back on itself the problem is much more obscure. The possibility of this chain relationship should be considered as giving a clearer understanding of the nature of a problem, even though it cannot actually solve the problem. Often, however, it may not be so important to discover too precisely which is cause and which is effect, provided that a causal relationship may be demonstrated. Life is full of ifs and buts. The whole of man's activity is a chained sequence of causes and effects. To attempt always to find a first cause is not practicable, but it is often possible and useful to trace the more important immediate-past causes at a stage when different causes (i.e. different action taken) might have given different results.

In the short-term prospect, it may be possible to focus one's attention on just one link in the chain and at that stage cause and effect may be distinguishable. Two possible simplified links in the economic chain might be:

1st — price increase reduces demand
2nd — reduced demand reduces price.

In the first stage, price change is the cause and demand-change is the effect, but the positions are reversed in the second stage. While the whole movement of price, demand, and supply is continuous and highly complex it is nevertheless made up of loops which, in certain circumstances, may be unlinked from the main chain, although at a later stage they may be re-attracted to the chain as if by magnetism.

One set of data alone will rarely, if ever, indicate the existence of a causal relationship; a mass of evidence will be needed for this purpose. Nevertheless a limited amount of information may reveal a degree of correlation between variables. Correlation occurs between two or more variables in circumstances which may fall short of permitting the expression of one as a mathematical function of the others. The variables are correlated if they behave in such a way that changes in the value of one are associated with changes in the value of the other and so that it is possible to predict the value of one variable if the value of the other variable is known.

Two variables which appear to be correlated may in fact have no significant connexion with each other except through their separate relationships to a third variable, just as a man's mother-in-law is related to him only because of her relationship to his wife. While one form of correlation between variables may suggest a cause and effect relationship between them, another may show a relationship between two variables which are themselves the effects of another cause. Two symptoms of a disease are related to each other only because of their individual relationships to the disease. Correlation must not, therefore, be confused with causal relationship but it may provide the pointers to the latter. One cause may have many effects, just as one apparently single effect may be the result of many causes, and the behaviour of the two known variables may assist in defining the behaviour of the third variable upon which they are both dependent.

Where no direct causal relationship can be established between two correlated variables, their correlation cannot be used as evidence to support the proof of any proposition other than their relationship to the variable through which they are related to each other. Changes in the price of beer and in the price of potatoes may be shown to be correlated in that rise trends in one are always accompanied by similar trends in the other. This is not necessarily because the price of beer is in some way dependent upon the price of potatoes, or vice versa. The correlation may merely be a double reflection of a general rise in retail prices of all goods. The correlation then exists because both variables are measured in terms of a third variable of money-value.

Apparent correlations can exist merely because of accidental numerical similarities between the values of variables within closed sets. The three blind mice and the three wise men have no point of similarity except that there were three of each of them; neither can they be related to each other by way of any other medium. Their sole relationship is that they form mathematical sets the members of which may be paired off so as to give a one-to-one correspondence; that is, there is one blind mouse to each wise man. If this one-to-one relationship was always maintained irrespective of the numbers involved, then by counting the men one could always tell how may mice there were. The sets are, however, limited to three members each and the circumstances limit the number of sets themselves to one each. It is certain that not each wise man has a blind mouse, however, and indeed it is not unlikely that there are more blind mice than there are wise men.

The term *correlation* has no practical meaning in circumstances such as these, yet in some similar circumstances mere numerical equivalence may be mistaken for it. Correlation coefficients are calculated mathematically from the available data in the form of numbers. A numerical coefficient may always be obtained by the appropriate procedure, but a numerical process can provide only a numerical answer. The latter then has to be translated back to reality from the mathematical model. Twice two will always give the answer of four in mathematics but this cannot be interpreted in every field of activity in the same way. If the tax on tobacco is doubled it does not follow that the total revenue derived from the tax will also be doubled, since less tobacco may be sold at the higher total price.

Mathematics operates in closed fields, although there are many gateways leading from one field to another. In expecting tobacco tax and revenue changes to be equivalent, one would be guilty of inaccurate translation – the wrong formula is being used. Other variables, such as the nature of demand for tobacco, have been omitted and the wrong formula can never provide the right result. Unfortunately, however, it may appear to do so and it is therefore essential that, in translating the significance of a correlation coefficient, the utmost care must be taken to ensure that all known factors and the possibility of the existence of unknown factors are taken into account. The most careful examination is necessary and common sense is as important here as it is in interpreting all other statistical results.

Correlation coefficients[1] are expressed in values ranging between -1 and $+1$. The nearer a value is to either of these extremes, the better is the correlation between the two variables. If the value is positive then the correlation is direct; as the independent variable increases, so does the dependent variable increase. If the value of the coefficient is negative then the correlation is inverse; as the independent variable increases, the dependent variable decreases. The closer the coefficient value is to zero, the less is the correlation between the variables.

Although the correlation between two variables may be established as being almost complete, it does not follow that the changes in the values of the variables are of equivalent absolute magnitude. The changes in these values will depend upon the near-functional ratio relationship between the variables. This may be demonstrated most easily by the behaviour of variables which have a complete functional relationship. If $y = 2x$, then y will be increased by 2 when x is increased by 1. The absolute variations in the respective values are not identical. The true relationship exists because the value of y is always equal to twice the value

1. See Appendix 1.

of x. This is a simple linear relationship so that it may be represented as a straight line. All the points plotting the paired values of the variables will fall on this line without deviation. If it were necessary to calculate the coefficient of correlation it would be found that it was equivalent to $+1$, since the correlation in a functional relationship is obviously complete. The coefficient may be said to measure how closely a correlation approaches to a linear functional relationship; that is, how close is the association between the two variables. It does not measure the quantitative ratio between the variables.

Comparatively few coefficients approach the limiting values except in the physical sciences or in man-made conditions where the relationships are first set up and then controlled. In circumstances where the uncontrolled elements of chance or human nature enter unseen into the equation, the correlation may be disturbed beyond recognition. It is always possible, however, that the recorded observation of two variables might have presented a different pattern but for the intrusion of these other factors. If the changes in values are of a major nature, so that the statistical correlation is low, then any real relationship between the two variables may be obscured. If, however, the statistical correlation is high, it is reasonable to assume that, despite the intrusion of other causes, the correlation may have a real significance.

To this end the corresponding values of the two variables may be plotted on a chart, and an attempt may be made to draw a regression line which would represent the linear equation to which the related behaviour of the two variables most closely approximates. The regression formula which decides the slope and axial interception of this line is calculated by the least-squares method formulated by the French mathematician Legendre. The line may be used to predict the approximate value of one variable, given a known value of the other. It can give only a prediction since the line will not pass through all or even any of the plotted points for actual observed values.

The regression line which gives the best fit for a set of observed values is that which, when drawn, is so placed between the points representing the observations that the sum of the squares of the separate distances between the line and each point is at a minimum value.[1] There are, in fact, two possible regression lines, one being measured against the x axis and one against the y axis.

The points charted in figure 23 are the actual observed values of the two variables x and y, whereas the line of regression predicts the values of x for given values of y, it being assumed that

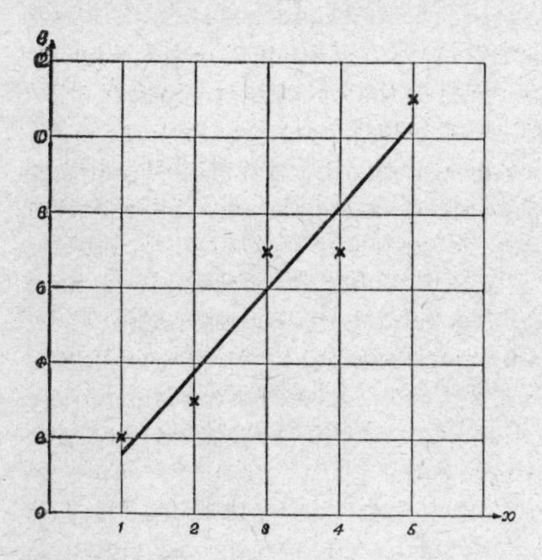

Fig. 23

Fitting of regression line to observed values of y (see Appendix 3 *for data)*

the equation for the regression line does in fact define an approximate relationship between the variables. Where the points are scattered closely about the line, so that the distances between the points and the line are short, there is a clearer possibility of the line actually defining a true relationship. If, however, the points are scattered all over the chart so that it looks like a close-up photograph of the measles, no line can be drawn to represent any relationship and it is absurd to make the attempt. Yet such attempts are often made and the regression line is sometimes presented on a chart which does not disclose the actual points. This is most misleading since it suggests that all the points actually fell on the regression line and this infers a complete functional relationship

1. See Appendix 3.

between the variables. A regression line, therefore, should never be accepted unless the observed values are also plotted on the chart.

If it is possible to draw a line which derives from some mathematical process, and to show that all the observed values approximate to it relatively closely, then the mathematical law may be used to calculate values for pairs of variables at levels which have not been observed. These values, being predictions, are necessarily approximations; the degree of their accuracy depends upon how well the line fits the observed values. There is, however, some danger in extending the line too far beyond actual observations. For certain values of x, the intruding factors may have only a slight disturbing effect upon the correlation of x and y, but for higher values the relative effect may alter. It is possible to have too much of a good thing. The growth of healthy bodies may, perhaps, be shown as related to the amount of exposure to the sun which the bodies receive, but too much exposure is definitely harmful. The relationship which was originally positive may have suddenly changed to a negative one.

Again, there may be limits to one variable or to both. The rate at which a tank fills up with water is related to the rate at which the water is released from the inlet pipe, but when the tank is full, not only does the level stop rising but also the flow of water into the tank ceases altogether. When an attempt is being made to predict the approximate values of a variable by regression methods it is therefore as well, if at all possible, to take actual observations for one set of paired values lower and one set higher than the level for which the prediction is sought. If each of these sets still approximates to the regression formula then, in the absence of any other contradictory evidence, it may reasonably be supposed that the desired values will also approximate to the formula – the limiting values do not appear to have been reached nor has the positive relationship changed to a negative one.

Whether an actual correlation exists or not is a question to be decided in each case separately. Much will depend upon the circumstances involved. There is, however, one particular fallacy which might usefully be noted here. This involves the assumption that a set of circumstances exists because of the existence of certain factors; and then, later, the existence of these factors is thought to be proved by the existence of the original circumstances. This circularity of argument will get one nowhere. If you start travelling eastwards around the world and continue ever onwards you will eventually return to the line of longitude from which you departed. You will start by moving eastwards but you will return from the west; east and west are only relative terms. Similarly the quantity A expressed as a function of the quantity B will express the relationship such that A and B vary together but it cannot prove whether A varies because B does or whether B varies because A does. It is therefore that much more absurd to say that A varies with B because B varies with A and then to 'prove' the latter by saying that A varies with B!

11 Artful Advertising

Published statistics often do not say what the readers take them to mean. When they are used in advertisements it is often impossible to decide what, if anything, they do mean. Many of the more obvious abuses of statistics in fact occur in advertising material. This, perhaps, is not really surprising; advertisers have their own particular axes to grind for otherwise they would not advertise. How often, for example, does one read an advertisement which says that a product is not worth buying? It should be said at once that not all advertising is so blatantly exaggerated as are the efforts of some of the less careful advertisers. To condemn all advertising because of the activities of the latter would be as illogical as trying to prove that all statistics are unreliable merely because some figures which are presented as true statistics are nothing of the sort. At the same time it cannot be denied that there has been a great deal of misleading material published in the advertising world.

This abuse is not, of course, confined to statistical representations; a misleading word inconspicuously inserted in a sentence may result in the reader making an interpretation which is quite unwarranted. Advertisers are entitled to their opinions. If they believe that their product is the best in its range, there is no reason why they should not say so, provided they make it clear that it is their opinion. This licence, however, does not extend to exaggerated statements in support of their belief; nor does it permit statements which, although true in what they say, omit other relevant details. An advertisement may be strictly true in what it says but untrue in what it does not say. Mental reservations may be as dishonest as absolute falsehoods.

Advertisers are themselves sometimes naïve in their claims

Bounteous eulogies are bestowed upon their products in a mounting crescendo of praise only to be followed by the assurance that their products have no existence in fact. At least, that is what they say and, in view of the properties claimed on behalf of the product, they may be nearer the truth than they suspect. But that is not what an advertiser means when he says that there is 'nothing like Blotto'. What he really means to say is that there is nothing *else* like Blotto. Whether or not this is true may be open to doubt, but at least the statement is more meaningful. The carelessness of the orginal statement leads to the paradox of an advertiser claiming that there is nothing like the product he describes and paying large sums to say so. Nevertheless this form of carelessness is not itself misleading. Other forms could be misleading, but quite apart from these it is regrettably true that the least one can say about some advertising material is that it is careless.

Peter Fleming[1] has referred to 'the lurid and terrifying regions of the advertisements. From every page a face stares out at you, haggard, tragic, haunted. The furrowed brows of spruce young business men who have been smoking the wrong cigarette, drinking the wrong kind of coffee, betray the imminence of a nervous breakdown.' He might also have referred to the 'before-and-after' conditions of those who have been rescued from such extremities. An advertisement, which claims almost miraculous curative properties for some remedy, may include the picture of a middle-aged lady who is suffering from indigestion and who is clearly in the worst state of depression. The accompanying picture of the same lady after the recommended treatment certainly reveals the miraculous powers of the nostrum. Quite obviously the sufferer's indigestion has disappeared but, then, so have the white streaks disappeared from her hair. Her hair has indeed become darker and wavy again, sunshine is now streaming through the window, and even the wallpaper is different. Wonderful indeed!

What has this to do with statistics? The before-and-after method does not involve numbers, but it does embody a technique which is basically statistical in that it seeks to compare measurements of some condition of well-being. The effect of the remedy

1. *One's Company* (Cape).

has been to change the measurable level of that condition in the patient. Before she took the remedy her well-being was below a particular level of measurement, whereas afterwards it was above that level. The measurements are not quoted but the fact of measurement is implied in the depicted difference. This is tantamount to saying that, in the particular unit of measurement selected, there has been a measurable change, and measurements are statistics.

Sometimes the advertisers will actually introduce numbers in their material by adding a caption such as 'Three times better on *Blotto*!' The pictures cannot, of course, depict the ratio of 3 to 1, but this difficulty is of little practical importance since no one can possibly know what the statement means. Three times better than what? Does one get three-times-better results from *Blotto* than one can obtain with the best competitive brand or with the average performance among other brands or, indeed, with the poorest quality alternative? Even if we could sort this out, how are we to measure betterment? What is the unit of measurement? Or perhaps the patient is supposed to be three times better now than she was before she took the remedy. But she was ill before taking the remedy. She was not well to begin with and she therefore cannot be three times better. It is impossible to compare negative and positive quantities in a positive ratio.

The truth is that the statement has no real inner meaning for the reader – or even for the advertiser – but the reader may not be looking past the statement to discover its true meaning. He may therefore interpret it carelessly. The advertiser, for his part, is not necessarily concerned about *how* the reader achieves his interpretation, for the statement can only be 'interpreted' in favour of the advertised goods. He is concerned only with the final interpretation, not the method by which it was reached, and as long as the reader can be brought to understand that the product is better than something, it does not necessarily matter to the advertiser what that something is.

Loose phraseology can work wonders. People do not expect to have to read advertisements closely word for word. Instead they are more likely to obtain a general impression of what appears to be the gist of an advertisement. They will readily absorb and memorize numbers without assessing their real significance; they will probably not read any qualifying words nor note the absence of relevant details that have been omitted. An electric heater may be advertised as the heater with the lowest running cost, but in what respect is it cheaper? Is the actual cost of operating the heater cheaper so that it will cost only so many pence to use it for a standard unit of time? This may well be true but it is not necessarily a virtue. A one-kilowatt fire, for example, will be cheaper to use than will a two-kilowatt fire, but the latter supplies more heat. If one ignores differences in electricity tariffs, it may perhaps be shown that, in theory at least, the cost per unit of heat produced is the same with both fires. The larger fire costs twice as much to use only because it is generating twice as much heat. Again, if a new house for sale is advertised as the cheapest on the market, it is fairly clear from the start that it will also probably be the smallest and that if you want a larger house you will have to pay more for it. The same applies to differences between fires having different output ratings. The likelihood of such differences should be apparent since there is a kilowatt output yardstick which may be used to relate the performance of individual fires.

Many articles, however, are described as being the cheapest in circumstances which the purchaser cannot check until after he has bought the articles. There are, for instance, many different models of electric shavers available at widely differing prices. Here the consumption of electricity is negligible. The efficiency of the instrument as a shaver is the true criterion. Cheapness of purchase and cheapness of operation are meaningless terms except as related to the performance or utility of the article purchased. A cheap tool is worthless if it will not do its work.

It should be noted here, of course, that some people can use tools better than can others, and that some products are more suitable for certain individuals than for others. The first differential may be overcome as the user becomes more proficient; the second differential may never be overcome for suitability cannot always be changed. The advertiser thinks of his potential customers as a group. The circulation of the newspaper he chooses will ensure that his advertisement stands a reasonable chance of being read by a calculable number of readers. If a certain percentage of this total of readers in fact buy his product, then his object is achieved. He wishes to sell, say, 10,000 articles and if he does this then he is content. He does not care at all whether Smith buys one or not since the personal identities of his customers are irrelevant.

Smith, however, looks at the advertisement in quite a different way. Wittingly or not, he looks for a direct personal relationship between himself and the manufacturer. He requires the article for his personal use and it must therefore measure up to his individual needs. The advertiser obviously cannot address his advertisements to Smith personally, even though he may appear to do so by sending circulars to his address. If Smith unquestioningly assumes that the advertisement's claims can be applied to his own particular requirements then some part of the misunderstanding is his own fault. He is making the mistake of taking a generalization, whether true or untrue, and applying it to a specific instance.

It now becomes clearer that even the kilowatt rating of a heater is not necessarily the correct criterion to use in assessing its relative costs. There are different types of heater with different uses. Some provide direct heat only within a limited distance, while others of the convector type spread the heat out in quite a different manner. Each has its own particular advantages and disadvantages qualifying it for its different uses. Instead, therefore, of calculating the cost of a stated output of heat (i.e. in terms of thermal units) one should perhaps consider the cost of the effective input of heat into the space being heated. Rated output and effective input of heat are not necessarily the same. Measurement of the input may indeed be extremely difficult if effectiveness is taken to include comfort or well-being derived by the purchaser from the heat obtained – this, after all, is the primary object in buying a heater. But because this cannot be measured exactly, it is not a valid alternative to accept comparisons of irrelevant details merely because the latter *are* measurable.

Advertisers are well aware of the difficulties of immeasurability. Cigarette manufacturers say that their cigarettes 'satisfy' or 'please', but they obviously do not mean that the cigarettes are guaranteed to please or satisfy every smoker. These terms are harmless enough in themselves since pleasure and satisfaction are relative words. Provided a smoker believes that he is deriving pleasure from a cigarette, then he *is* deriving pleasure for this is implicit in his belief. Similarly, a purchaser can easily tell when he is warm, but he has to become warm first and the kilowatt rating of a heater may not help him to judge the exact warming effect of the heater. Nevertheless this rating is, perhaps, his only guide. If, having bought a heater, he finds that it does not suit his purpose, he will be very ready to allege that the advertisement was misleading, but it may be that he has been led astray by his own limited knowledge as to the significance of kilowatt ratings.

The advertisement may therefore have been unintentionally misleading. This is not always so. If an advertiser cannot claim a specific virtue for his product he may nevertheless be able to produce some apparent evidence which proves something else altogether. By an association of ideas the product advertised may be given the appearance of possessing a non-existent virtue. A tyre manufacturer, for example, may head his advertisement: 'Worn Tyres are Dangerous!' This is an arresting headline and the advertiser may follow it up by saying that his tyres do not wear out so quickly as do those of other manufacturers. Both facts may be true, but neither supports the other. Worn tyres are always dangerous, whatever the make. The real point of the advertisement is that the particular tyres do not wear out so quickly, but the reader may also read into it the statement that they are also safer. In the absence of evidence in support of this safety aspect, such an inference is invalid.

The advertisement may also quote irrelevant statistical data in support of the headline. For instance, the headline might read '1,651 more accidents last year'. The accuracy of the data may be undoubted and its exactness adds its own air of respectability, but it has no direct reference to the advertised durability of tyres. It proves only that there have been more accidents; the condition of the tyres of the vehicles involved may have played no part in the increase. There may have been more accidents because there were more inebriated pedestrians or simply because there were more cars on the road. To argue, from the evidence available, that the increase in the number of accidents was in some way related to the use of worn tyres is as absurd as claiming that, since there are more new cars in use there must also therefore be more new tyres in use and that, as accidents have increased in proportion, then the increase in the number of accidents is the result of the use of an increased number of new tyres!

The last example involved the use of legitimate but irrelevant statistics, but statistics may be so manipulated that not only are they irrelevant, they are downright illegitimate. This is made possible by the device of the shifting definition.[1] A cigarette manufacturer who has tested his brand against others may report that 3 out of 10 smokers considered his brand was 'freshest' to the taste and that 5 out of 10 considered them 'smoothest'. The meanings of these two qualities are not made clear but the fact that they are separated indicates that there is meant to be a difference. The manufacturer then says that 3 out of 10 judgements were favourable to his brand in one category and that 5 out of 10 were favourable in the other category. He concludes therefore that 8 out of 10 smokers credit his brand in one or other of the categories of freshness or smoothness.

This conclusion, however, is true only if the individuals who judged the brand as freshest did not also judge them as smoothest. It is true only if those in favour in either category are mutually exclusive. If a smoker finds that a cigarette is both the freshest and the smoothest available, then he is represented in both categories so that, when the category totals are added, he will in effect be included twice. This would be clearer if the proportions were:

Smoothest:	8 out of 10 men
Freshest:	5 out of 10 men

Adding these together would at once show up the fallacy since it would appear to produce a proportion represented as 13 out of 10. It would be equally misleading to say that there were 13 smokers out of 20 who judged the cigarette favourably, because the five men who found the cigarette freshest might also be included in the eight smokers who found it smoothest.

This example deals with a method whereby an advertiser could seek to prove that his product was more popular than the figures

1. See Chapter 8.

really revealed. At the other end of the scale is the advertiser who seeks to show that, in a range of products which may have a harmful content, his product is least harmful of all. Cigarettes, for example, are frequently suspected of being the cause of this or that contemporary complaint, usually with very little evidence. But although the scientists may not be able to establish a causal relationship, they can analyse the constituent parts of the tobacco and the smoke given off. If various brands are all analysed in exactly the same way it is fairly certain that not all brands will score identically similar results as to the amount of poison content. This being so, it will be possible to place the brands in a ranking order according to the respective amounts of this poison. One brand will show up as having the highest whereas some other brand will show up as having the lowest poison content. The manufacturer of this latter brand then advertises that his cigarettes are better than others in that the statistics proved that they had less poison content. He does not quote the actual statistics for a very good reason. In the nature of things the amount of poison in cigarettes is likely to vary within very narrow limits as between one brand and another. The actual difference may be so small that it is in fact hardly a difference at all.

The size of an advertised difference should be revealed. In some ways of life the very difference itself is of significance irrespective of the size of the difference. In athletics, for example, four runners may cross the finishing line in a race within fractions of a second of each other but the closeness of the result is of no value in competitive sport. The athlete who first crosses the line is declared the winner irrespective of the fact that the next man was so close behind him that, in terms of actual running performance, there was virtually no difference between them. This may be hard on the second man but it is all part of the accepted risks of the game. In statistics, however, a difference has no significance in itself. A difference is not a difference at all unless it really matters.

Quite apart from this distinction between mathematical and significant differences, there is another fallacy which has crept into the cigarette manufacturer's argument. In effect he is saying that there is less poison in his brand of cigarettes and that therefore they are also safer. This inference does not necessarily follow from the facts. If one cigarette contains enough poison to cause harm, the fact that another has a higher poison content may not mean that a relatively greater harm may be caused by it. The harm is already done with the smaller dose. If it takes x ounces of arsenic[1] to cause death, then two separate quantities of x and $2x$ ounces would each cause death even though one quantity was only half as great as the other. It is, in any case, an odd thing to advertise that one's products contain poison!

Advertisers are, of course, pretty well obsessed with the 'more and better' tradition and, although the percentage sign is not always used in this connexion, it is ever present lurking in the background and is at its most persuasive in advertising. Statements which include comparative terms like these should always make it clear just what is being compared. 'Ten per cent more people are buying . . .' run the headlines. Ten per cent more than what? Does this mean that more people are buying now than they did a year ago or just a week ago? This could be an important distinction where there are seasonal fluctuations in the demand for a product. Does the claim perhaps mean that more people are buying more often or does it merely mean that sales have increased by 10% and that it is therefore assumed that the number of customers has increased in the same proportion? Whatever the intended meaning, it can never be clearly appreciated by the reader unless he is also told to what base figure the 10% proportion applies.

Even more dubious is the use of phrases such as 'ten per cent more goodness in . . .' for they bring with them even wider difficulties of interpretation. Again it is necessary to know against what the manufacturer is measuring his own product, but the reader also wishes to know how the manufacturer has calculated his 'goodness' ratings. Is he comparing the goodness to be derived from his product with the goodness that can be obtained from a potato or a vitamin pill; has he taken into account the relative bulk and costs of the items compared; or is he merely saying that his own product now has ten per cent more goodness than it used to have, thus inferring that it used not to be so good

1. This example is not connected with cigarettes.

as it might have been? One cannot be sure. All that the reader can be certain about is that the statement as it stands is meaningless!

In athletics only those competitors who enter a race may be compared in relation to their performances. If they do not enter a race they obviously cannot win it. This is not confined to sport where such limitations are voluntary. Other supposedly wider categories are arbitrarily restricted. The Wholesale Clothing Manufacturers' Federation has the pleasant fancy of selecting the ten best-dressed men in Britain, but they can select the winners only from the men whom they have inspected and these are normally very much in the public eye. Just because the Federation selects them as being the best-dressed, it does not mean that they are really or necessarily the best-dressed men at all; they are merely the best-dressed ones who have come to the notice of the Federation. In much the same way a Beauty Queen is not necessarily the most beautiful woman in the country, though she may be the most beautiful amongst those who entered the contest. If it is possible to decide what one means by most beautiful and to rank all ladies so that the most beautiful one could be identified, it might well be that that particular lady would refuse to enter the contest. It is always necessary that the reader should be quite clear as to what subjects are being compared and as to whether the comparison may be accepted as being related to a total population or merely to a sample.

Reported proportional differences need particularly close scrutiny when they originate from sample surveys. 'Housewives say that colds clear up eleven per cent more quickly with . . .' claims the advertiser. The first thing to note here is that the housewives have probably said nothing of the sort; they are more likely to have submitted figures to some collecting agency which has then calculated some form of average proportion. The next point to note is that no matter how accurately this figure may summarize the results obtained, the results themselves must have originated from a sample survey since clearly not every housewife could have been interviewed. The accuracy of the final statistic will be subject to normal sampling errors and, in the event, may be subject to serious bias if the survey has not been conducted with due regard to statistical requirements. A third important point is that the advertisement mentions only housewives. Are housewives the only people who catch colds? This is important because it shows that doubt exists as to whether they were in fact reporting on their own colds or whether they were acting as reporters for their whole families. If the first alternative applies, then the sample is not representative of the whole population of cold-sufferers. If the second alternative applies then, in order to relate the results from different homes, it will be necessary to weight the individual results so as to take account of different family sizes.

Then, having taken these points into consideration, what is the real meaning of the statement that colds clear up eleven per cent more quickly? It could merely mean that colds clear up in eight days instead of nine. This is hardly a triumph for a cold remedy. In any case, however, some definition of a cold is required and, further, there is the problem of how one decides when a cold has actually cleared up since it may leave traces of catarrh and other discomforts for many days. With all these doubts and qualifications in mind, the reader may be excused for thinking that the apparently exact statistic of 11% begins to take on that spurious air which has been identified elsewhere.

The presentation of statistical or pseudo-statistical information in advertisements is enormously assisted by the use of charts. It has already been noted[1] that an illustration of facts often has a greater impact than the facts themselves. Examples from the whole range of pictorial illusions have been employed at one time or another in advertisements but the latter have also developed their own particular forms. It is simple to exaggerate the apparent significance of a line or other kind of chart merely by changing the scale of the drawing. Some advertisers in the past have apparently found it difficult to decide either what would be the correct scale to use or what would be the best scale to produce just the line that they wanted; and they have overcome this difficulty by drawing the line on a grid without any scale at all. Such a 'chart' is of course as meaningless as a dictionary without words. The variable whose progress is thus plotted (perhaps in

1. Chapter 3.

more senses than one) is in the same exposed condition as a ship whose navigation charts are blank; it may finish up anywhere.

The before-and-after technique may also be incorporated in this connexion by using two different scales so that the 'after' chart appears to reflect some significant development. Figure 24 shows such a possible combination.

The gradient in the second chart looks impressive and the reader may be tempted to believe that it must mean something.

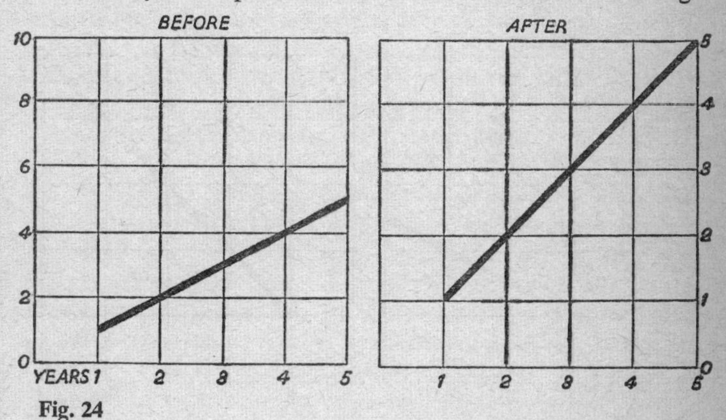

Fig. 24

There is nothing, however, to show that the scales are identical; the complete charts might be as in figure 25.

These charts reveal that the data charted are identical but there is a world of difference in their appearances. It will also be noted that none of the charts indicates the exact nature of the variable supposedly represented. This is art for art's sake. It is the picture which counts all the time, not the sitter. Yet it is a paradox that some forms of pictorial representation may be so exaggerated that they no longer become misleading. An example of this is given by an advertisement (issued by the Guinness Brewery Company), wherein a man is shown bending an iron girder into the shape of the numeral 5.

This company has had honourable connexions with statistics[1]

1. See *Journal of the Royal Statistical Society* (*Series A*), Part 1, 1960, pp. 2–7.

and one would not expect them to tamper with statistical propriety. Instead they have produced a number of amusing advertisements which succeed in communicating numerical information to the general public. It is true that they started with an advantage in that the figure they wished to represent – 5 millions – was so great that it could hardly be exaggerated in the public mind; anything over a million is pretty large by ordinary standards. Numerical exaggeration was therefore next to impossible and

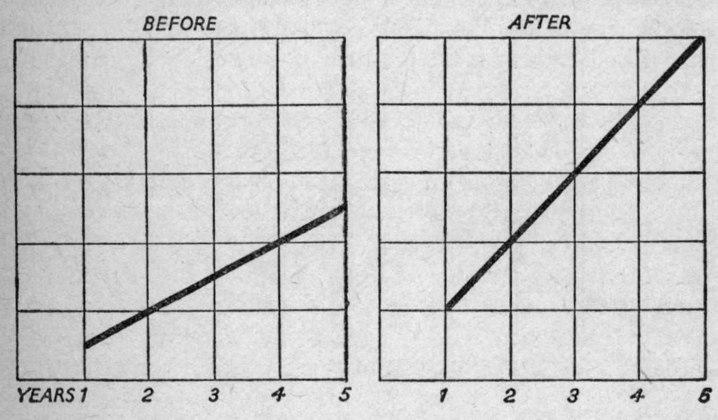

Fig. 25

the rounding off of the figure to the nearest million ensured that no one could mistake it for an exact figure.

The success of the picture lies in the fact that it represents a complete descent into fantasy. No one expects fiction to be true, nor true stories to be fictitious. A fictitious illustration of a fact is therefore permissible only when it is possible to distinguish between the fact and its representative fiction. Nobody believes or even expects that a man will be able to bend iron girders with his little fingers and will not therefore be misled into believing that a regular dose of Guinness will enable him to do so.

But between fact and fancy lies the intervening obscure perplexity of shadows where truth becomes compromised by fiction. Historical fiction, for instance, is based upon fact but distorts reality in such a way that, except to a student of history, it is often impossible to distinguish the one from the other. Statistical representation sometimes suffers in the same way.

The abuses mentioned here are not, of course, confined to advertisements. They occur in statistical representations of all forms of activity where the reader or the perpetrator is insufficiently informed on the methods of statistics. The advertising world however, can scarcely deny that they have by usage developed an advertiser's licence in this respect. Advertisements rarely portray real life at all any more than do other similarly glamorized forms of literature. Life is not all misery for the man with indigestion nor is it all smiles for the fortunate purchaser of the sovereign remedy.

There is really only one way in which to test the true worth of a product and that is to test it for oneself. Although this may not always be practicable, particularly when the product is expensive, the reader should never accept advertised statistics as in any way guaranteeing the same satisfactory results as could be derived in an actual personal test. He should ignore the statistics unless they stand up to the scrutiny which they deserve and even then they should be accepted only as pointers to information and not as substitutes for the information itself.

12 Is there an Index?

Index numbers are calculated in order to provide in a single term an indication of the variation against time of a group of related values. They are intended to show variations in magnitude which are not susceptible to direct measurement or observation. Entries in an index to a book give a ready reference to particular aspects of the book but not to the book as a whole. An index number also points to a particular aspect of a group of variables, but it refers to the group as a whole.

If it is desired to compare changes in the price levels of two commodities over a number of years, so as to show how one has moved in relation to the other, it is usual to calculate price relatives. A specific year is selected as the starting or base year and the commodity prices in that year are equated to 100%. Prices in later years are then expressed, within the commodity groupings, as percentages of the base year prices.

Figure 26 shows changes in the prices of two commodities. The price of commodity A has increased over five years from £50 to £61 per ton. The latter price is 122% of the former. Similarly the price of commodity B in Year 5 is 150% of its price in Year 1. In Year 5, therefore, the price relatives of A and B are 122 and 150 respectively, the percentage signs being dispensed with. These relatives facilitate the comparison of price movements of commodities without much fear of misunderstanding and this method may be used for comparing changes in the values of any variables such as the volume of production, sales turnover, and others. It is often possible to produce useful answers to the question of to what extent has the value of A increased or decreased in relation to the increase or decrease in the value of B. The only possible misinterpretation of simple relatives is that, because they are expressed as relatives, then the variables measured are correlated to each other. This is not true. The only relationships revealed by these relatives are between the individual values within their own sets and not with values in other sets. The relatives may, however, be used to demonstrate a correlation between variables if a change in the value of one is always or usually matched by a change in the value of the other.

Price Comparison

Year	*Commodity A*		*Commodity B*	
	Price per ton (£)	*% on Year 1*	*Price per ton (£)*	*% on Year 1*
1	50	100	48	100
2	51	102	49	102
3	53	106	52	107
4	59	118	60	125
5	61	122	72	150

Percentages calculated to nearest whole number

Fig. 26

Just as it is sometimes useful to compare movements in values of simple variables, it is also extremely useful to be able to track general movements in the values of more complex variables consisting of whole groups of items, such as the retail price level of all commodities or of wholesale prices in general or within certain specified manufacturing categories. This is where the index number offers its services. It is in fact a particular form of average of the relatives of the simple variables which together comprise the complex variable, and it seeks to indicate general movements in the latter.

This is asking a lot of a single number. In considering general trends of retail prices it will be found that the price of one commodity may rise while the price of another may fall and the prices of the various commodities may all react in differing degrees. How can one index number satisfactorily represent all these changes? The answer, bluntly, is that it cannot and is not intended to do so. It is not intended that the index number shall give any indication of changes in the values of the individual variables, but instead its purpose is to average the nett effect of all the changes.

The Index of Retail Prices is the best-known index in Britain. Before the usefulness of this index may be examined it is perhaps advisable to give an outline of its structure.[1] During a period

Commodity Groups included in the Index of Retail Prices

1 Food
2 Alcoholic Drink
3 Tobacco
4 Housing
5 Fuel and Light
6 Durable Household Goods
7 Clothing and Footwear
8 Transport and Vehicles
9 Miscellaneous Goods
10 Services

Fig. 27a

of twelve months beginning towards the end of January 1953, some thirteen thousand households kept records of all their expenditure. This information was processed by the Ministry of Labour and National Service, being summarized under the main group headings as shown in figure 27*a* and further subdivided into sections within the groups. The food group, for example, is divided into sections for flour, butter, beef, and other food products.

Altogether there are ninety-one sections which originally were so formed as to include every main item of expenditure covered by the index. Certain payments shown by householders were excluded altogether – such as income tax and life insurance payments, subscriptions, gambling expenditure, doctor's fees, mortgage payments, and others because of the non-measurable nature of the services acquired in return for the payments made. It was also impossible to identify a unit of expenditure for these items

1. *Method of Construction and Calculation of the Index of Retail Prices*, H.M.S.O., 1956.

which would retain its identity and so be capable of measurement from time to time.

Since it would also be impracticable to measure the price changes of every single item purchased, a selection was made of those particular items in each section which could be considered to be representative of the section. Thus, in the bread section of the food group, the following items were selected for pricing:

$1\frac{3}{4}$ lb. loaf — white
14 oz. loaf — white
14 oz. loaf — brown

There are, of course, many other varieties of bread, fancy or otherwise, but the theory is based on the assumption that if one type of bread becomes dearer then all other kinds of bread should also become dearer in proportion. Altogether some 350 commodities and services are priced regularly. Individual price relatives are calculated (with prices as at 7 January 1956 as the bases) and these relatives are then averaged to give the final index number.

The emphasis on the original collected data was not so much on the prices charged by suppliers as on the amount of money expended for each household to a standard averaged pattern derived from the survey. Not unexpectedly it was found that most households spent much more money on bread than they did on fish. A variation in the price of bread would therefore have a greater effect upon total expenditure than would an equivalent variation in the price of fish. To take account of this difference of effect, each separate percentage change is weighted in accordance with its relative importance to the total expenditure before being included in the averaging calculation.

A notional example of this calculation is shown in figure 27*b*. The total in the product column (2,670) is divided by the total number of weights (89) to give the average price rises (30%) for the food group. This is the increase on the price as at the base date; the base price is taken as 100% and the index number for this group is therefore 130.

It was claimed that the sample used for the survey was representative of nearly nine-tenths of all households in the United

Kingdom, giving a picture of the average expenditure pattern of all households of wage-earners and most households of small or medium salary earners. The Ministry also made it clear that the index was to be an index of price changes and not an index of changes in the cost of living. This disclaimer was advanced because of the lack of precision in defining the cost of living, but there is no doubt that, whatever definition is used, the most important factor in the cost of living is the level of retail prices.

Calculation of Index Number

Item	*% rise in price over base date*	*Weight*	*Producer*
Bread	31	19	589
Flour	29	4	116
Beef	17	24	408
Pork	20	6	120
Fish	31	9	279
Butter	34	12	408
Tea	50	15	750
Total	—	89	2,670

Fig. 27b

Does this index now really reflect the true changes in retail prices? Nobody would expect it to be exact, but it is nevertheless a serious attempt to trace general movements. Much loose talk is used in connexion with subjects like the cost of living and it is useful to have some method of approximation whereby one may obtain an idea of reality. How well can an index perform this function? The answer to this question is that it can probably serve the purpose, provided we do not expect too much of it. There are many opposing views on the validity of the principles employed as well as on the usefulness of individual indices for specific purposes. There are many difficulties involved and, indeed, the problem is so complex that one must admire the attempt to solve it even if reserving some doubts as to the value of the results achieved. Something attempted might well be something achieved, but this aphorism is subject to a number of qualifications.

Some errors are bound to creep into a series of calculations so involved as those required in the computation of retail price index numbers. So many averages and approximations form the basic data used at different stages in the process and there are also variations in the treatment of the data. The prices which are included are those actually charged for cash purchases without noting any discounts allowable unless they are generally available. No allowances are made, however, for Co-operative Society dividends which effectively reduce prices, but the prices of goods sold by retailers at cut-price levels are included at the lower rates. This is a fine distinction; the cut-price retailer charges less at the time of purchase whereas the Co-operative gives the money back later.

Some differing methods of pricing are used for different produce groups but, in general, a price index is calculated for each item within a product group for a number of selected towns. These indices are then tabulated into population groups and an index is averaged for each population group. The resulting five index numbers are again averaged to give an index number for the product item for the whole country. The actual collection of the basic price data, however, has its own particular difficulties. Some items of expenditure do not form easily recognizable units; expenditure on holidays, on education, and on hobbies, among others, all come within this category and, as a result of the original survey, certain arbitrary decisions had to be made as to the best way in which to deal with these more elusive indeterminates. Whether these decisions were correct is a matter of opinion; some of them are referred to in this chapter.

Many questions may be asked as to the efficiency of this index. The pattern of expenditure on which it is based was derived originally from a representative sample of households, but were the records accurate in the first instance? In this connexion, there is a surprising admission in the official notes.[1] For a small

1. *Method of Construction and Calculation of the Index of Retail Prices*, H.M.S.O., 1956.

number of items, recorded figures were adjusted 'because there was evidence that the expenditure on these items had not been fully recorded'. These items were in respect of alcoholic drinks, tobacco, confectionery, and similar comforts and the recorded figures did not agree with other available statistics which were considered to be more reliable as indicators to expenditure levels for these items. Recorded figures which disagreed with other data were rejected, but recorded figures for which no other 'reliable' statistics were available were accepted. One cannot help but entertain some disquiet on this point for there is an implication that, had other statistics been available to refute the recorded details, then the latter would have been rejected. This does not suggest any great measure of confidence in the accuracy of the recorded details which were used to form the framework of the index.

It will be noted that, if the accepted external statistics were really reliable, the items upon which the sample households were reticent included those little personal luxuries the details of which one member of a household might wish to conceal from the other members. A husband with a large alcoholic intake might wish to minimize this in his report in order to avoid trouble with his wife. But if he took a shilling or two from the expenditure on beer, would he not also have to add the same amount on to some other item of expenditure, possibly bus or rail fares, in order to account for his total expenditure? If he did in fact do so then these other affected items of expenditure were also unreliable. There is no way of identifying these affected items, but the mere fact that this effect was not taken into account must upset the true balance of weighting employed for the average calculation.

A much more serious defect of the index, however, is that the pattern of expenditure, whether perfectly recorded or not at the time of the original survey, is not a static pattern throughout time. The pattern itself will change as new fashions and inventions change our way of life and as different products, from time to time and for a number of reasons, become generally preferable to others. The overall pattern will also change in association with changes in the ratio between the number of adults and the number of teenagers in the total population of a country. Recent years have seen remarkable increases in expenditure upon gramophone records; the prices of these records have also fluctuated within fairly wide limits and yet record prices are not included in the index at all. The accepted pattern of expenditure was laid down in 1953–4 but the index was not commenced until 1956. The latter year was taken as the base year for the calculations even though the expenditure pattern might well already have altered by then; it has certainly altered appreciably now.

The individual expenditure pattern for a particular household, and therefore indirectly for the whole population, depends upon the spender exercising his choice among various possible purchases within his income capacity. He must choose the nature of his outlay. Prices for all commodities do not rise at the same rate and one kind of purchase may become more attractive than another. A rise in the price of one commodity may be sufficient to preclude its purchase by a householder or, if it is an essential item, its purchase may mean that he cannot buy some other less necessary commodity for the simple reason that he cannot afford to do so. Whether they are direct substitutes for each other or not, all items offered to an individual for the favour of his selection compete for the amount of money he has available. Prices themselves will therefore have an effect upon changes in the pattern of purchases. Even if the prices of all commodities rose in exactly the same proportion, the pattern of expenditure would probably still change. A householder could not accommodate a ten per cent rise in the retail price of every item of his normal purchasing pattern unless his income also rose sufficiently to meet the increased costs of his purchases. Income increases do not necessarily match increases in prices except in those industries which have contracted agreements stipulating such a relationship.

The pattern of expenditure is, of course, of primary importance to the whole index. What of the prices themselves? Within each pricing section certain commodities only are priced as being representative of the whole section. This is probably a safe device for bread prices. Account is taken of changes in prices of ordinary white bread and brown bread, considered separately, and these together probably do reflect major changes in the price of bread generally. That the same device is a generally satisfactory method for all sections is not so certain. In Section 74 (writing paper and other stationery goods) only writing paper and ink are priced. Is it really true that the prices of the majority of goods sold by stationers follow the price-changes of these two items?

It may well be that these two items are the only stationery goods of importance to the members of the households included in the original survey. This serves to emphasize the important fact that the index is restricted to considering trends in the retail prices of goods normally purchased by households which, at the time of the survey, excluded those in which the head of the household had a weekly income of £20 or over. The published title of the index is therefore not sufficiently precise. It does not even seek to measure retail prices of goods which were outside the buying range of the households surveyed.

The theory on which an index is based is that measurements of the same variable should be recorded from time to time for the purpose of comparison and that the measurements should then reveal the trend in the values of the variable. The essence of this is that identical things should be measured. This is not as simple as it sounds, since it presupposes the absence of changes other than price change. Prices do not change in a vacuum; the character of the thing measured may also change. The nominal price of a product may remain unchanged whereas the quality or the unit of quantity may change its identity. To compare the two prices would not then be realistic since the two products measured would not be identical. A better quality article would be expected to be more expensive than a similar article of relatively poorer quality. The amount of soap powder in a packet may vary from time to time although the size and price of the packaged goods may remain the same. This is a real change concealed by an apparent similarity.

There is also a possibility that an article, which was originally selected as representative of its section, may in fact be displaced as the true representative by some other article as a result of the mutations of consumer preference. To price an article, the use of which has decreased so that it is now used less than an article which is not priced, cannot give a truly comparative result unless, by chance, changes in the prices of both articles are closely correlated.

In mathematics the device of using indices (as in x^2, x^3, x^4 . . . etc.) enables the numbers so represented to be related to each other because they can all be expressed in terms of the same base number x. The base number remains identically the same. A price index would be equally precise if the representative articles remained identical. The world, however, is constantly changing. If we calculate the mean height of a family of four children and then, a year later, if we calculate a new mean height, it would be absurd to compare these two means as if the measurements were identical in character. The first mean represents the height of children of a particular age group; the second represents children in a different age group since they are all one year older. The children are the same in their family relationship to each other, yet not the same.

This changeability of the index structural pattern is an accepted statistical risk. The index cannot give perfect results but it does make an attempt to obtain approximate measures of something which is not strictly measurable. The rise in retail prices is brought about by a complex structure of cause and effect in which system and accident alternate. Retail prices reflect the whole economic life of a nation and, although the index is specifically a price index, it is impossible to dissociate the buyer from the seller. They are both parties to one transaction. The retail price to the seller is the same as the cost of living to the buyer. General retail prices and living costs are two very generalized aspects of one immeasurable quantity. Yet it is worth noting that approximate measurements of different aspects of what is apparently the same quantity will not always agree. 'Measured by expenditure data, the gross domestic product is estimated to have been 3½% higher in the third quarter of 1959 than in the same period a year earlier. Based on income data, equally valid estimates in the change in the gross domestic product show an increase of 5½%.'[1] A difference is bound to arise since the calculations are based on different data which, because of the arbitrary nature of some of the definitions

1. *Economic Trends*, H.M.S.O., February 1960.

within their respective indices, are not representations of identical things.

To define what is meant by income and expenditure respectively is possibly so complicated that one cannot be expressed as the exact opposite of the other. Yet they may both be employed in index calculations with the confusing results admitted in the above statement, and the reader is left asking himself which is correct. By 'correct' he will mean which adequately reflects reality, and in this context it is sad to have to say that probably neither is correct. The true result may lie somewhere between the values quoted and although there is no guarantee of this, the best one can say is that the change is approximately between 3½ and 5½%. The result from each index is in any case only an approximation and it may be that the two indices taken together, although seemingly confounding each other, could instead set up notional limits of error for each other.

It is a paradox of the index system that, because of its limitations, small mathematical changes are not really significant whereas greater changes, which should also be relatively more significant, can usually be shown only over a long period during which time the basic expenditure pattern is almost certain to have altered. Yet in the long term retail prices do move appreciably. Everyone can watch certain items becoming dearer and find that a pound sterling gradually buys less and less. Some general measurement is certainly desirable and one must use whatever measuring instrument is to hand. An approximation is all that is required, but is the approximation good enough? Some form of checking is required to watch this point.

The real check is carried out in respect of the original pattern of expenditure since this is the beginning of the race. Regular surveys are needed at the end of each lap to make sure that the runners are the same competitors as those who were in at the start. Another device would be to commence at the end of a lap and to trace the runners' footsteps backwards; that is, to fix the present pattern and to trace price movements of the components of expenditure revealed. This may, of course, merely reverse the snags since purchasers will today be buying articles which were not even on the market ten years ago, but it would have the advantage of dealing with the position as it really exists now. We are concerned with the past only as a means whereby to measure the present. It is more realistic to say that the articles now in use are twice as dear as they were so many years ago, than it is to say that articles which were once in use would now be twice as dear if they were still used.

Such a procedure would not have been possible in the past. A survey of this nature has always been a prolonged affair because it has taken so long to process the data. The original survey for the Index of Retail Prices was begun in January 1953 but it was not until 1955 that certain of the information was made available to the committee responsible for formulating recommendations. Until comparatively recent years it has rarely been possible to publish results of large-scale surveys until well after the events they recorded. To attempt to discover something about the present it is not of immediate practical use if the answer will not be known for two years or so. The increasing use of electronic computers, however, makes it possible to process data much more rapidly and it is to be hoped that up-to-date checks and revisions of indices may become established practice.

We have so far discussed only one index number system, but all systems of this nature suffer from the same defects. In 1959 the United Commercial Bank[1] called for a revision in the base for the index of industrial production in India. The series of index numbers at the time was based upon the values for 1951 but since that date many new industries had developed. The expansion in these new industries had not unnaturally been more rapid than in others, yet because of their lowly beginnings in 1951 they had been accorded relatively low weightings. The index therefore failed to measure most of this expansion.

In the United States an index of industrial production is produced by the Federal Reserve Board. This index, with 1947 as the base year, revealed a level of production in June 1958 which was 55% higher than in 1947. A new index with a new structure based on 1957 production data showed[2] the increase over the same period as 66%. About one-third of the difference resulted

1. *Fortnightly Business Letter*, 14 December 1959.
2. *Economist*, 9 January 1960.

from the widening of the scope of the index beyond manufacturing and mining to bring in utilities such as electricity. The other two-thirds difference was due to improvements in individual series, the introduction of new series, and the reweighting of component parts of the index to take account of the relative values of net output in 1957.

It is another paradox of index numbers that the structure of the series needs to be fairly rigid so that it remains basically the same throughout, yet it also needs to be flexible so as to take account of changes outside itself. Changes in the index structure must not be of too high a degree, for then the revised series in reality becomes quite a different series instead of a continuation in a different form of the former series. Many index numbers, when published, are accompanied by conversion factors relating the present series back to previous scrics and therefore to different basic conditions. Such conversion factors should be used with some caution since there is not a true relationship between an original and a revised series. Clearly, if the original series was satisfactory in itself, then there would have been no need for the revision.

The Index of Retail Prices, for example, has 1956 as the base year. If it is desired to relate price changes to earlier years, this is not possible by direct methods since the index was not then in existence. There was in existence, however, an interim index for which the level of prices at 17 June 1947 was taken as base. These two indices differ in their scope and also in their methods of calculation, but it was deemed possible to relate them approximately by the conversion factor $\frac{153 \cdot 4}{100}$ since the interim index number was 153·4 when the new index was started at 100. Thus, the index number at 12 January 1960 was 109·9 based on 1956 figures, but was equivalent to

$$109 \cdot 9 \times \frac{153 \cdot 4}{100} = 168 \cdot 6$$

based on 1947 figures.[1] This, again, is only an approximation, but

1. *Method of Construction and Calculation of the Index of Retail Prices*, H.M.S.O., 1956.

it was necessary to establish some form of relationship because of the existence of agreements in certain trades whereby wage levels were determined in relation to movements of the interim index.

To meet some of the difficulties, a form of chain or moving base index has been devised. The idea, briefly, is to derive the index number for any year by using the previous year as base instead of relating every year to just one base. The advantages claimed are that weighting may be changed from year to year, that new variables may be included as required and that it is generally possible to keep the index series up to date. At the same time, however, the amount of work involved in this would be immense and the too frequent alterations in the structure of the index might conceivably destroy its usefulness except in the very short term.

Index numbers of different series are often called into use to assess the degree of association between the variables represented. Thus it is possible to compare movements in retail prices with movements in wage rates by comparing the trends of their index numbers. Oddly enough it is because the two series measure obviously different variables that they may be used together. Extreme care, however, is essential in attempting to compare two series of apparently similar variables. To compare the production index number of the United Kingdom with the production index number for India, for instance, it is first necessary to ensure that both series have the same general structure. The nature of the goods produced would obviously differ since different countries have their own special products, but it would also be necessary to discover what omissions were made from either index and to ascertain what differences, if any, there were in weighting methods or in the collection of the data or in its processing. Index numbers of the cost of living might vary largely as between different countries merely because of the differing concepts of what constituted an average standard of living.

The direct comparison of percentages derived from different indices, without regard to the underlying details of construction, provides the kind of circumstances in which the persuasive percentage revels. The latter is also prominent in other calculations involved in the computation of index numbers. The final number may depend upon the averaging of percentages from group series which are themselves the results of averaging the percentages derived from sections within the groups. The averaging of percentages is a dubious process unless due regard is given to the relative magnitudes of the values of the variables in each set of data for which the percentages are calculated. This difficulty is easily overcome by a suitable weighting procedure. Percentages can, however, appear to behave most oddly even where there are no apparent weighting or structural problems, as is shown by the following example.

The prices of meat and potatoes as at 31 December in two different years were as follows:

	Last year	*This year*
Meat	6d. per unit	1s. per unit
Potatoes	1s. per unit	6d. per unit

Meat is now twice as dear as it was per unit, while the price of potatoes has been halved. It is instructive to observe what happens if these figures are combined into an index series. Has the index number become higher or lower as a result of the changes? If last year is taken as the base year, the calculations give the following result.

	Last year	*This year*
Meat Relative	100	200
Potato ,,	100	50
Totals	200	250
Combined Index	100	125

The combined index, which is the average of the two individual relatives, shows that on average prices have risen 25%.

Perhaps this sounds an exaggerated figure? Then, we will take this year as base instead. This will give quite a different result, as follows:

	Last year	*This year*
Meat Relative	50	100
Potato ,,	200	100
Total	250	200
Combined Index	125	100

This shows that the average price has decreased by 25%. By changing the base year, we have changed the result from a positive to a negative one. One method shows that prices have gone up, whereas the other method shows that they have gone down. Which is correct? In this particular instance it may be noted that, if one purchases one unit of meat and one unit of potatoes in each year, then neither answer is correct since the total cost is 1s. 6d. in both years. There has been no change in total at all! The same index relationship would, however, be derived from the following data:

	Last year	*This year*
Meat per unit	2s.	4s.
Potatoes per unit	1s.	6d.

There clearly is a movement here since what originally cost three shillings in total now costs four shillings and sixpence. That is to say that present prices are now 50% higher than last year.

There is, of course, a fallacy in all this. It lies in the fact that the percentage relatives within the meat group and those within the potato group are not percentages based on the same quantity. A one-hundred per cent increase in the price of meat would represent two shillings where the same percentage increase in potato prices would represent only one shilling. The respective percentage changes must first be weighted in proportion to their relative base prices. The weighted percentages are thus expressed to a common base and may now be mathematically combined for averaging purposes.

The calculation for the last example, taking last year as the base, then becomes

	Last year	*This year*	% *relative*	*Weight*	*Product*
Meat price (in pence)	24	48	200	2	400
Potato ,, ,, ,,	12	6	50	1	50
			Totals	3	450

The average product, 150, is the combined index number and shows that prices have increased by 50%.

If, instead, we take *this year* as base, the calculations are

	This year	*Last year*	% *relative*	*Weight*	*Product*
Meat price (in pence)	48	24	50	8	400
Potato ,, ,, ,,	6	12	200	1	200
			Totals	9	600

The average product is $66\frac{2}{3}$ which is the index number for last year. The index number for this year (as base) is 100, so that the increase is $100 - 66\frac{2}{3}$. The increase between the years is therefore $33\frac{1}{3}$ which is exactly 50% of last year's index number. This again shows that prices have risen 50% since last year. The same answer is derived from both calculations.

The fallacy of the shifting base does not in fact exist in a well-ordered index but it is apt to appear in unofficial indices where the compilers are not so careful. A well-ordered index is one which satisfies the time-reversal test which is embodied above. This is based on a criterion proposed by Fisher that a good index satisfies the relation.

$$\frac{x}{100} \times \frac{y}{100} = 1$$

where x and y are the separate numbers calculated with the first and second years as base years respectively. The amended methods above do conform to this criterion for

$$\frac{150}{100} \times \frac{66\frac{2}{3}}{100} = 1$$

The arithmetic mean, however, will only satisfy this criterion if the price relatives are first weighted. An unweighted index is not an index at all.

Most important indices use the arithmetic mean as the appropriate statistic; some, however, use the geometric mean instead. There are two main reasons advanced in favour of the geometric mean, neither of which stands up successfully to close scrutiny. One reason given is that the geometric mean is less susceptible to major variations as a result of violent fluctuations in the values of the individual items. The *Financial Times* industrial ordinary share index employs the geometric mean 'since for practical purposes only a small number of shares are included and they cover a wide range of industries so that a movement in one is not necessarily indicative of the state of the equity markets as a whole'.[1] The adoption of the geometric mean in these circumstances is therefore largely a matter of expediency to overcome an inherent defect arising from the small number of items included in the index. No method, however, can really overcome the defect that an individual movement may have little relation to the field surveyed as a whole. This particular shortcoming can only be remedied by making the component items of the index truly representative of the whole field; otherwise the index numbers derived cannot be representative. The damping down of the effect of one unrepresentative item does not by itself inspire greater confidence in the accuracy of the final statistic.

The other main reason advanced in favour of the geometric mean is that it is mathematically more suitable. It is true that the geometric index always satisfies the time-reversal test,[2] but it is an illusion to believe that it is therefore correct. Conformity

1. *Financial Times*, 5 March 1960.

2. See Appendix 4.

with this test is an attribute of a good index; but it does not follow that an index which conforms is therefore necessarily a good index. The reversal test is satisfied in the calculation shown in figure 28 but the result is not what might be expected.

Geometric Mean and Time-Reversal Test

Item	*Last year*	*This year*	*Relatives*	
			This year as base	*Last year as base*
A	10	5	200	50
B	4	10	40	250
	Products of relatives		$8{,}000 = y^2$	$12{,}500 = x^2$
	Geometric mean (approx.)		$89{\cdot}4 = y$	$111{\cdot}8 = x$

Fig. 28

The reversal test is satisfied since

$$xy = \sqrt{12{,}500 \times 8{,}000} = 10{,}000$$

$$\therefore \frac{xy}{100 \times 100} = \frac{10{,}000}{100 \times 100} = 1$$

The result shows that, with last year as base, the index has moved to 111·8 which is equivalent to saying that the values have increased by 11·8%. In fact, however, the actual increase in total values has been from 14 to 15 so that this increase is one-fourteenth of the base figure. This is equivalent to 7·1% and does not agree with the geometric mean.

The only really satisfactory mathematical treatment of the subject is to use the weighted arithmetic mean. Nevertheless, the basic prices and the relative price-changes in the above examples are so different that, in practice, it would be meaningless to combine them in any average form of statistic. Any so-called index would entirely conceal the basic details. This may be an obvious example of unsuitability of method, but such divergences may perhaps occur on a smaller scale and yet not be noticed in a more complicated index structure.

13 Time Series and Arithmancy[1]

A time series is simply a set of recordings of the values of variable quantities measured at intervals of time. The basic quality of the data from which such series are derived is one of historical fact. The values are records of actual measurements which have already taken place; the instant a measurement is made, so it becomes part of history. The interest in tabulating and charting such a series may therefore be purely historical, or it may be concerned with trying to establish possible forms of correlation between different variables, or it may be even more concerned in predicting probable future events.

Time-series methods have been criticized for this one main reason – that they reflect a living in the past. Look to the future is the slogan. Yet, as soon as it is attempted to give a time series a futuristic look, the criticism becomes even more emphatic. Is this criticism justified? The historical value of a series is real enough. In commercial life, for instance, the managing director of a company will want to compare last month's sales with those for other periods and to know which products appear to be gaining or losing ground. He will also want to test the efficiency of his company's various other functions – what is the accuracy of the budgeting procedures; did the introduction of new methods have the expected effects; does there appear to be a general change in the industry's affairs as a whole?

All these facts and many more may be brought out by the proper tabulation and charting of data. It is, admittedly, all history now, but it is not necessarily lacking in significance merely because of that. Mankind would be greatly the wiser if it could

1. Divination by numbers.

but learn how to make the fullest use of the correct interpretation of history. History can be applied to the future in a limited way. The danger is that one may not read it correctly. To ascribe specific results to wrong causes is as bad as ignoring history altogether. History in the long term and on a large scale is extremely complex. It is often impossible to find a first cause and sometimes difficult to find a level at which a different course of action would have had different results. In the short term, however, and within more restricted limits of application, it is possible

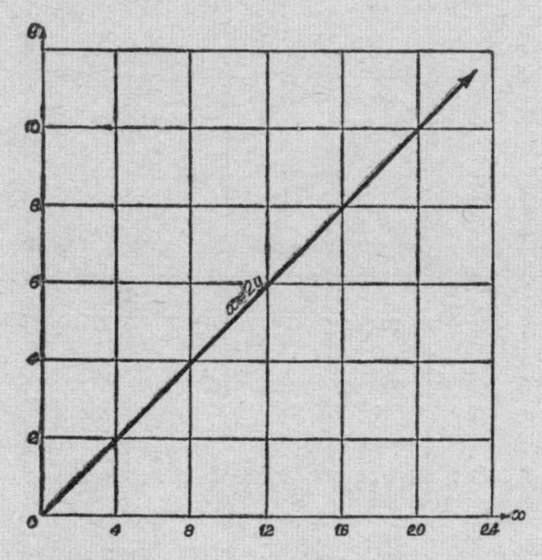

Fig. 29

to trace the effects of certain causes. It is not enough to know that something has happened. It is also necessary to have some idea as to why it happened and what conditions made the occurrence possible. Mistakes or successes once recognized can perhaps be avoided or repeated, as the case may be, provided they are fully understood. It is odd that political history should be accorded such an important role in the academic world and yet that there should be a body of opinion which decries historical statistics as worthless.

Possibly the criticism, which surrounds any attempts to extend a series into the future, arises because so many attempts are ill-founded and carried out without due regard to common sense.

Such criticism is fully justified, for attempts at forecasting are often based insecurely upon a confusion between the character of an ordinary time-series line and of an algebraic curve. A mathematical curve, which is derived from an equation in which one variable is consistently represented as a fixed function of another, will have a definite shape either as a smooth curve or as a straight line as in figure 29.

Any and every point on the curve gives pairs of values which

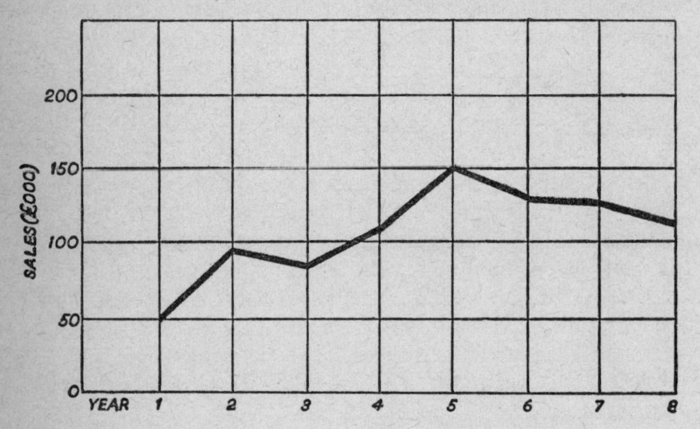

Fig. 30

satisfy the equation. A time series, however, is merely a series of recorded measurements which do not form a series in the same sense. Its line which connects the points representing the values will not have a regular shape, but will vary with the data fluctuation so as to resemble alpine scenery as in figure 30.

The line is nothing more than a means of joining a number of otherwise unconnected points. The sectors of the line between each two date points have no meaning with regard to intermediate dates. This fact is probably quite clear in figure 30 but, when the line becomes less irregular in movement so that it approaches the resemblance of a mathematical curve, it sometimes leads the reader astray into believing that it is a mathematical curve. It is possible to derive values direct from a mathematical curve by interpolation even though physical measurement has not been made. Thus, in figure 29 the curve may be drawn by joining only two points ($x = 10$; $y = 5$; and $x = 20$; $y = 10$). All other pairs of values may then be read off the curve (e.g. when $x = 8$, $y = 4$). In the same way values may be extrapolated by extending the curve since the further direction of the curve, although not actually drawn, is nevertheless known. A time-series line, however, has no recognizable shape and, since it is merely a stringing together of measurements, it is not possible to derive other values from the line itself. The values charted have no mathematical relationship except that they are allegedly all measurements of the same variable.

Statistics, as historical records, illustrate the effects of certain causes whether known or unknown. They have no intrinsic value in themselves and are of value only as guides to the causes and effects which they reflect. The numbers alone cannot therefore be associated mathematically to predict future measurements of the same variable. Nevertheless the causes and effects which they reflect may, when identified, be applied to forecasting probable future results.

Forecasts of the future size of a national population may be predicted with sufficient confidence to provide the base for a number of calculations involved in problems affecting the economic needs of that population. The projections of population and age distribution are essential aids to educational authorities, local and national government departments, and companies whose activities involve large capital investments and whose plans must necessarily be formulated in advance – often a number of years in advance of the anticipated demand for services and commodities. 'In recent years, Canada has experienced a more rapid increase in the size of the population than most countries. Since the last complete census in 1951 recorded the total population, it has risen at a compound rate of close to $2\frac{3}{4}\%$ per annum as a result of high birth-rates, declining mortality, and large-scale immigration.'[1]

Information such as this is invaluable in predicting future

1. Bank of Montreal *Business Review*, 23 February 1960.

trends, but the forecaster has to keep a very close watch on significant changes in the present which will affect his forecasts for the future. The population bulge in the United Kingdom, which came as a result of an abnormally high post-war birth-rate, has been used to great effect by some commercial minds who, probing ahead, realized that there would be a correspondingly high demand for products required for children, and that this

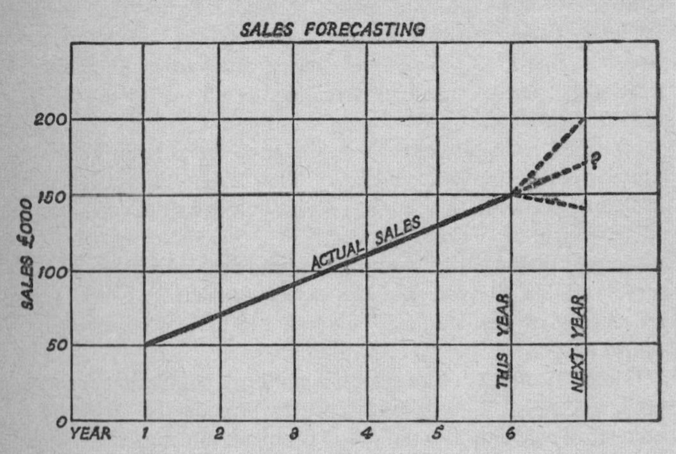

Fig. 31

would be followed later by a quickened demand for teenagers' requirements. Changes in the age distribution are as important as changes in total population, since they can produce wide changes in expenditure patterns.

A time-series line may be almost exactly straight. This may, however, be purely accidental unless there is a reason for a steady and regular change in one direction. There is no justification in assuming that the extension of the line in the future will continue in the same direction. The causes must first be analysed and the likelihood of their continuance must be assessed before any attempt may be made to project the line into the future.

Figure 31 shows a straight-line representation of a time series. Up to the present date the values recorded are definitely known and the line is thickened. How will the line progress in the future? This can only be estimated; some estimates will be better than others but the fact that they are estimates should be shown clearly by dotted lines.

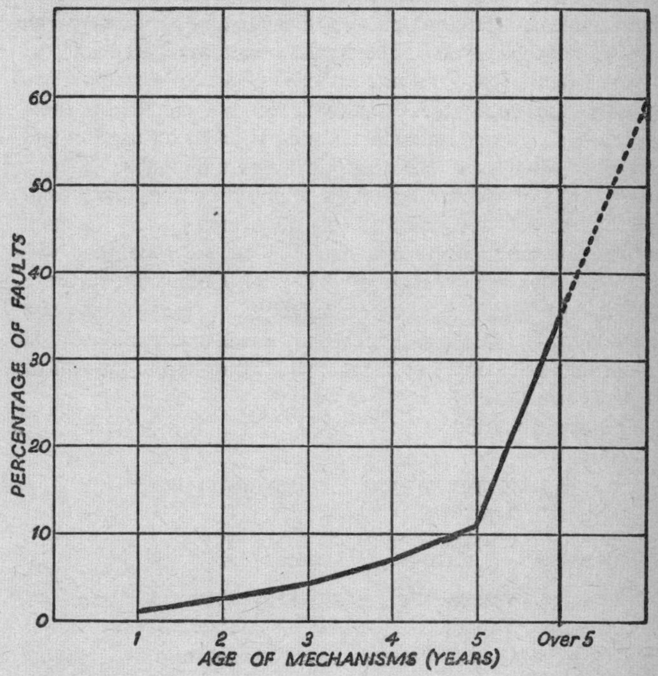

Fig. 32

Everyone likes the game of forecasting in some form or another, whether it is betting on races, playing the stock markets, or merely forecasting the constitution of representative football or cricket teams. To this extent, therefore, most so-called forecasts are coloured by personal experience, opinions, and prejudices, and they will often finally depend largely upon hunches. This cannot be disputed, but the use of historical data, by revealing trends and cycles in different forms of activity may, when used in conjunction with considerations of statistical probability, at least help to point in the approximate direction which the line should follow.

Forecasting future trends solely from the use of historical data, without first divining their true meaning, is like the favourite party blindfold game of pinning the tail on the donkey. The direction of the line cannot be deduced from the chart itself; but it very often is! A novel and almost unbelievable example has recently come to light. A chart (figure 32) was prepared to show the incidence of reportable faults in a particular mechanism and sought to establish a relationship between the number of faults and the age of the mechanisms in which they were discovered. The number of faulty mechanisms discovered among the total number of mechanisms manufactured in any one year was represented as a percentage of that total. Complete data were available for the mechanisms manufactured in each of the years from 1955 to 1959, but figures for mechanisms manufactured before that date could not be segregated according to the year of manufacture.

The points charted were:

Age of mechanism	*Fault percentage*
1 year	1%
2 years	3%
3 years	5%
4 years	7%
5 years	11%
Over 5 years	35%

The number of faulty mechanisms in the first year was negligible but increased to 11% among mechanisms which were five years old. The chart then shows a tremendous increase in faults, but the steepness of the line's gradient is seriously in error. The first five classes along the x axis (age of mechanism) each refer to one year. The sixth class, however, refers to an unstated number of years; it might refer to a period of up to twenty years. The true rate of increase of faults with age is therefore distorted out of recognition by the fact that, although the classes along the axis are not equivalent, they are nevertheless accorded equivalent sectors of the axis. The figure of 35% is an average for all annual classes included in the 'over 5 years' class and does not measure the same variable as do the other percentage values.

That, however, is not the oddest part of the chart. The stated classes together include all mechanisms of all ages whatever; those which are not shown in their actual year of manufacture are nevertheless grouped together in the 'over 5 years' class. Every possible mechanism was therefore represented, albeit incorrectly, on the chart. Yet the chart, as drawn, so bemused the artist with its apparent significance that he was persuaded to extend the line into a further section on the assumption that faults occurred more rapidly as age increased. What this section was thought to have represented not even the draughtsman who drew it could have comprehended. Nevertheless the chart was solemnly published as being of importance.

The tendency to pin tails on to line charts is a common affliction. The chart of past and present national production rates of a consumer durable may show a steady increase over a period of four years, but it does not follow that the same rate of increase will be maintained for ever or even for the fifth year. The demand for consumer durables is not a simple function of any one factor. What is the productivity of the country's economy; can this be maintained or increased so that purchasing power may also be increased; might some new invention do away with the necessity for a particular product; will government policies be changed and so affect hire-purchase arrangements?

Even if the progress shown by the chart should be continued at the same rate, it would not follow that an individual manufacturer's chart for his own products could be drawn parallel to it. Some firms will progress more rapidly than others. The more economically efficient manufacturer could produce more units at lower cost and perhaps increase his own share of the total market, and other manufacturers might enter the market. This does not mean, however, that no attempt should be made. The modern gearing of large-scale production requires that next year's figures should be estimated within limits. In the event, one has to take a chance that uncalculated factors may intervene to upset a budget, but the wise man will accept as many pointers as are available to him.

This has been described as analytical guesswork and there is a deal of truth in this. Even if the answer is wrong, perhaps there is some virtue in one's having been helped to avoid making the same mistake again! Actual results should be measured closely against the forecasts. The mere fact that these agree numerically is not sufficient to prove that the forecasts were right. Pupils taking mathematics examinations will be awarded no marks if they accidentally achieve the right results by incorrect methods. The company cashier, who always makes his final cash balance agree with his forecast, may earn some repute for himself in this respect by the simple expedient of not paying any of the creditors' accounts. It is not an advantage to the company if the cash balance is forecast accurately at the expense of upsetting the creditors.

Whether or not statistical methods may be applied to forecasting future related events, they may certainly be applied to the interpretation of past events. It is possible, for instance, to ascertain whether the data reveal a trend in the movement of the values or whether their occurrences are erratic. If there is a trend, does this disappear and recur over the years so that the possible existence of trade cycles may be investigated? In the short term, there may well be evidence of seasonal fluctuations. To some extent the existence and general measurements of these movements will be sensed without recourse to statistical analysis. Everyone suspects that the sales of ice-cream decrease in quantity during the winter so that the manufacturer will have to adapt this production to different levels of demand at different times of the year. Again, it is general knowledge that the number of cars registered for licence purposes is increasing and the statistics would be expected to reveal this upward trend.

The manufacturers need more precise information than this. It is not enough to know that ice-cream sales are lower in January than they are in August. They will also want to know by how much the sales differ and will need to have more precise definitions of the periods of fluctuation. It is comforting to a motor-car manufacturer to know that sales of cars are rising, but he will also want to know how many extra cars were sold last year or, preferably, last week. The ice-cream manufacturer knows intuitively that there is a seasonal fluctuation in his trade but he will look for the existence of any evidence which may point to the development of an upward or downward trend in general over the year as a whole. More ice-cream is sold in August than in January, but he will also desire to see data which will show whether there is any major difference between the sales figure in one January and sales in the previous January. Is ice-cream becoming more or less popular? Unless he looks for possible trends, he will not know of their existence and may well be caught unprepared for changes in the size of the total demand. When the sales decrease in, say, October, he may accept the decrease as a seasonal drop but he should closely watch the extent of the drop. The decrease might be more than enough to account for normal seasonal fluctuations. The normal levels should be marked with danger signals just as the limits of safe sea-bathing areas are marked by warning notices. It is potentially dangerous to proceed beyond them.

A trend line may be fitted to a time series by means of the least-squares method[1] described in connexion with correlation methods, provided the data justify a linear trend. This method has been justified in that if there is a trend at all it can only exist because of the implication in its definition that there is a correlation between time and the variables measured.

Compound interest payable upon a loan increases year by year because it is calculated in such a way that it shall be so related to time; interest payments are deliberately tied to the calendar and future payment details are known because they have been fixed in advance. Other variables may conform to the same approximate pattern without actually being tied to it in the same way. Their relationship to time may be merely accidental or it may be that changes in their values are indirectly related to the progress of time in that their growth is a function of time in the same way as is the physical growth of a plant. Time is not the cause of the growth, but the growth may be measured against time in a recognizable pattern and is therefore closely related.

1. See Appendix 3.

The analogy of physical growth is a useful one because normal development involves a growth upon previous growth in the same way as one sector of a trend line continues on from a previous section. It differs in that growth upon growth is a compounding process whereas the trend-line extension merely carries on as a projection of the original line. Nevertheless the analogy is useful and is most meaningful in regard to the limitations of the application of time-series methods. The growth of a plant or of an animal develops upon itself until a certain state is reached when it slows down and then stops altogether. Nothing will grow for ever. Similarly, a trend line cannot always continue in the same direction. All sets of data have their numerical limits, whether the scope within the limits is known or unknown.

The trend will eventually flatten out or fall away at some stage in time and the earliest possible warning of such an occurrence is clearly desirable. There is, therefore, a need for the establishing of warning limits, but before these can be posted at the appropriate levels, the effects of yearly seasonal fluctuations must be considered. This is a worthwhile exercise, for a full appreciation of these effects is essential if one is to look beyond them to the underlying basic movement of the line. Changes in trends are likely to be hidden and disguised between the peaks and valleys of seasonal variations unless something positive is done to locate and identify them.

The most usual method employed in the adjustment of series is that of calculating ratios to moving averages. This first requires that an idea of the periodicity of the seasonal cycle shall be obtained from a number of previous investigations. If it is found that there is a clear-within-year cycle, as for example in the number of marriages registered, then the first step is to calculate what percentage of the yearly total of values is, on the average, represented by each statistical period (e.g. week or month) within a year. If there are twelve such periods in a year and each accounts for one-twelfth of the total annual values, then there is no seasonal fluctuation. If, however, it is found that, say, January on average accounts for only one-twentieth of the total, then some adjustment is necessary. Since, if there had been no fluctuation, the month would have accounted for one-twelfth and since in fact it accounts for only one-twentieth, it follows that if it is desired to forecast the year's figures from the January figures alone, one would need to multiply them by $\frac{20}{12}$ (or 1·67 approx.) to equate them to the average monthly figure for the year as a whole. Each month's figures would be treated in a similar manner, thus providing a relative factor for each month so that, when the

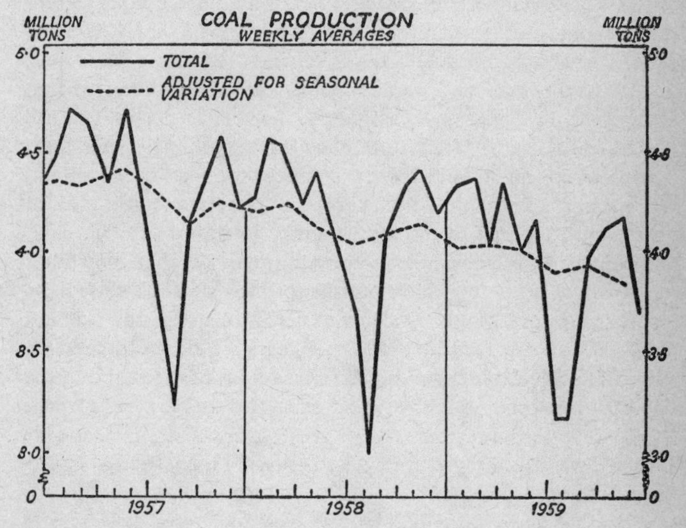

Fig. 33

actual value for each separate month is multiplied by its relative factor, the result in each case is equivalent to the overall monthly average for the whole year.

The results arising from these multiplications are the adjusted values and these are charted instead of the actual values to provide the adjusted line. It is usual to include both this line and the line which plots the original values on one chart, in order that the significances of the lines may be interpreted together. An example of such a chart is given in figure 33.[1]

This adjustment procedure flattens out the peaks and the dips.

1. *Economic Trends*, H.M.S.O., January 1960.

If all the actual between-month differences are entirely due to seasonal fluctuations then the adjusted line will be perfectly horizontal. This result is, of course, most improbable because of outside factors which are not apparently due to seasonal fluctuations and also because of some minor defects in the method itself. At the other extreme, it might be that the unjustified elimination of pseudo-seasonal movements could dangerously iron out significant changes which might not then be recognized until it was too late to take appropriate action.

Some refinements are required in the application of the method described. Calendar months vary in length. Thus, for the full calendar month of January 1960, which was allotted one-twelfth of the total annual values as its standard proportion, we should in fact allow 31/366ths to take account of the number of days in the month relative to the total number of days in the year. If, however, production statistics are being analysed, then only those days in the year on which production was possible should be included. Thus it would probably be necessary to exclude Sundays and the standard allowance for January 1960 would then be given as the ratio of the number of weekdays in January to the number of weekdays in the whole year. This would be equivalent to 26/314ths, and to forecast the year's figures from the January figures, we would therefore have to multiply the latter by a ratio equivalent to

$$20 \times \tfrac{26}{314} = \text{approx. } 1{\cdot}656$$

This ratio would still need adjusting for holidays.

Other points which must be carefully considered cover the fact that no one year is exactly like another, either in total or in relative monthly performance throughout the year. Other difficulties arise, particularly in production and similar statistics, because Easter is a movable feast and the holidays sometimes fall in March, sometimes in April. The resultant decrease in production figures, as a result of the holidays, does not always occur in the same period of every year. The same difficulty obviously arises in respect of the Whit holiday. Very little can be done about these variances, but they need to be considered objectively when data are being interpreted.

The calculation of seasonal fluctuations cannot be precise and it is therefore possible that nominally adjusted data may still retain some seasonal effects. The degree of accuracy in calculating fluctuations depends largely upon the regularity of their occurrence in point of time and upon the consistency of the degree of fluctuations from the basic movement. Some seasonal fluctuations entirely derive from or are amplified or reduced by the state of the weather in addition to mere calendar effects. So-called seasonal fluctuations of this type may not commence or even occur at all in identical calendar periods every year. Fluctuations in the sale of ice-cream will occur in the summer whatever the weather, but if the weather is abnormally hot in an earlier month, a greater than the expected degree of fluctuation may occur before the main season has really commenced.

In circumstances such as these, fluctuations and trends are apt to become confused. How can one tell if an apparent pre-seasonal fluctuation is in fact the beginning of an expected fluctuation occurring earlier than usual or whether it is evidence of a new trend? It may not be possible to decide this from the available data and the appearance of the movement should be treated with caution. The manufacturer should, however, place himself as nearly as possible in the desirable condition of being ready for any development. Background information is essential. Every salesman will know if his sales are reduced long before the final statistics are ready for analysis and he will have a good idea why this state of affairs should have developed. Their reports should be analysed and read in conjunction with the purely statistical results.

Because of the inequality of month-lengths, it has become the practice in many costing and other statistical applications to divide the year into thirteen periods each of four weeks' duration. This does not overcome the difficulties posed by the Easter and Whit movements and is not really a great advance upon the division of a year into twelve monthly periods. It does, however, have some limited advantage in that the basic periods are nominally equal (although adjustments have to be made to the first and last periods in each year), and that every period can be arranged to commence on the same respective day of the week, that is to say, on a Monday. Calendar months do not have identical days of the week relative to the days in the month. The first day of January and the first day of February, for instance, can never *both* fall on a Monday.

Provided the many limitations are borne in mind, a useful form of chart in connexion with time series is as given in figure 34.

The chart is divided into two sections. The left-hand section shows annual totals of sales for previous years. The section on

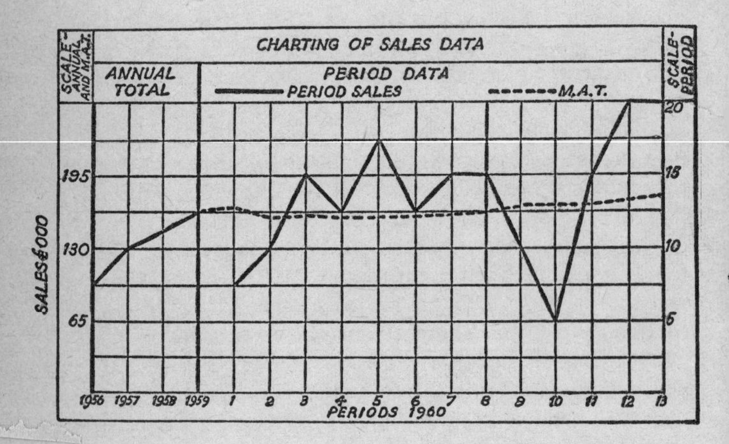

Fig. 34

the right shows separate results for each statistical period of the current year and also shows moving annual totals. Two different scales are used in order that the lines may be encompassed within a comparatively small area. There are thirteen statistical periods in each year and the overall period average is therefore one-thirteenth of the annual total. Accordingly it is convenient and useful for the ratio between the two scales to be thirteen to one. This has the advantage that, while the yearly totals and moving annual totals charted are totals for full twelve-monthly spans, the same lines also show period averages for those years if read against the period scale.

Actual period results may therefore be compared direct with the average figures for preceding calendar years and also for the last full twelve months. The moving annual totals also make it possible for period figures in one calendar year to be compared with the figures for the same period in the previous calendar year. The moving annual total at any one period is the total for the thirteen consecutive periods consisting of that period and the twelve preceding periods. Thus, at the end of Period 3, the moving year would be from the beginning of Period 4 in the previous calendar year up to the end of Period 3 in the present year. At the end of each calendar year the total for that year and the moving annual total will be identical because at that time the calendar year and the moving year coincide. The moving annual total at the end of any other period is calculated by adding the figure for that period to the previously existing moving annual total and then by deducting the figures for the same period in the previous year.

If the total sales for Year 1 are £343 thousand then the same figure represents the moving annual total as at 31 December. The moving annual total for the first period of Year 2 is then:

Previous M.A.T. (£000) =	343
Add First Period of Year 2 =	37
	380
Deduct First Period of Year 1 =	30
	350

It will be seen from this that differences between relative periods will be revealed by the movement of the moving annual total line. The above calculation merely increases the moving annual total by the difference between the figures for the separate periods. The moving annual total thus serves two purposes. It shows, for each set of thirteen consecutive periods, how the total is changing and this may be directly compared with the totals for calendar years; it also reveals differences between the figures for relative periods in two consecutive calendar years. This is most useful as there is thus no need to superimpose on the chart another line representing actual period results for the previous year; these may be deduced from the moving annual total.

The moving annual total line, if read on the lower scale, also gives values for period averages over thirteen periods and this serves to place periodic fluctuations in their right perspective. A manufacturer's sales of some products, for example, fluctuate fairly widely as between one period and another, quite apart from the effect of seasonal changes. The line therefore helps to give a sense of proportion to the whole chart. Restrained confidence in a continuously rising moving annual total, or concern at a consistently falling one, are emotions more justified than the alternate ecstasy and panic which could result from studying isolated periodic results.

The moving annual total gives an idea of between-year trends since, if sales are increasing in every period, as compared with the same periods in the previous calendar year, the line will continue to rise. It is not a trend line in the sense that it illustrates an approximating equivalence of the values charted to some mathematical formula. Nevertheless, it does, within its limitations, reveal trend movements. It is most unlikely that the line will ever be perfectly horizontal or even a straight line throughout the whole of its length. It is itself subject to minor fluctuations; sometimes it will rise and sometimes it will fall, but any real trend will become visually apparent.

If there were any valid basis for a high correlation between calendar time and other variables (e.g. sales) charted, then a trend line might be fitted by the least-squares method.[1] This, however, is not always justified. A straight line is much more likely to be more fiction than fact and has the disadvantage that it appears to point significantly in a definite direction. The invitation to extend it into the future is often too strong to resist. The gently fluctuating moving annual total line, however, feeling its way cautiously over the squared paper, lacks this appearance of spurious certainty and is therefore a more reliable guide to reality.

1. See page 319.

14 Probability

Probability theory figures largely in statistics. Some sets of data do not point to definite or even to approximately accurate results, but if one asks instead what is the *probable* result of certain courses based upon the interpretation of the data then it may be possible to provide a satisfactory answer. There are many quantities which we would like to measure but which, for a variety of reasons, are actually immeasurable. Where so much is uncertain because we cannot undertake the actual measuring, it is necessary to devise some means whereby we can measure the probability that inferences about the immeasurable quantities are correct.

By definition a finite population is of definite size. There is some value which represents the total of the population but the nature of the population may be such that it is not practicable to evaluate its exact total by actual measurement. How many households possess an electric washing machine? At any fixed point in time there is a definite value for the total number of such households, but it could never be accurately calculated without checking upon every household in the country. This would be an impracticable proposition. The probability that a selected cross-section of the population might in fact be representative of the whole population has opened the way to a great deal of statistical investigation which would otherwise have been impossible.

Probability is a measure of uncertainty. Where everything is certain there is no room for the generally accepted idea of probability since no doubt exists. Plato declared that 'arguments derived from probabilities are idle' but he was referring, not to mathematical probability, but to the popular intuitive ideas of probability. These everyday ideas are indeed too vague for analytical purposes. A man may venture an opinion that a particular outcome of an event is 'quite likely' or 'probably true' or 'not really very likely'. If these answers are given after due consideration of the problem, the speaker may have a mental image of the degree of probability involved but his answers completely fail to pass on to others a clear representation of this idea. In the beginning of man's counting habits, primitive tribes used the terms of *one*, *two*, and *many* as their numbers or they described objects either as being big or small. Modern mathematics could never have been developed on this meagre scale of measurement and, in much the same way, statistical inference requires that we must first have a means not only of discerning a general shape of probability but also of measuring its dimensions.

Measurement implies the observation of certain rules. The French philosopher J. Bertrand once asked[1] 'how can we venture to speak of the laws of chance? Is not chance the antithesis of all law?' To say that the outcome of an event is determined by chance is to admit that we have no idea how it is determined. The word *chance* is thus a collective one embracing all the unknown causes as well as all those circumstances which may not be the direct results of causes at all and which may be classified as mere coincidents. Nevertheless, even though we may not know the identities of the chance components, we can measure the results or coincidents and if these appear to form some kind of general pattern, then we may be justified in accepting that pattern as an approximately satisfactory working basis. It is doubtful whether domestic cats realize that their milk comes from cows, but they do not refuse to drink it merely because they are ignorant of its origin.

Aristotle once said that the probable was that which usually happened, but it must not be forgotten that the improbable does sometimes happen. The scientific mind will reject the idea of luck insofar as this defies mathematical analysis, yet this may be based upon a very fine distinction indeed. If a man tosses some coins then the probability of certain results being achieved may be calculated from the results he obtains in his experiments, the unknown factors which operate to produce these results being

1. *Calcul des probabilités*, 1889.

attributed to chance. Chance is a synonym for these unknown factors and this is very much what the ordinary person means by luck. Yet there do sometimes appear to be exceptionally lucky people who are always winning sweepstakes and enjoying other fortunate circumstances. This appearance may more often arise from intuitive judgements and, if the results over a long period were studied fully, it might be found that the lucky persons did not always win after all.

Nevertheless, if it is shown that over a period their results really are different from those achieved by other people, it would be unwise to ignore this difference. Perhaps they have found some way of eliminating one or more of the supposedly unknown factors so that they may achieve results which do not conform to the general pattern. This does not mean that the original probability calculation for the whole population is wrong, since the calculation applied to a set of circumstances different from those which apparently apply in the exceptional cases – the individuals involved in the latter are not members of the population of ordinary individuals.

It may be opportune here to note an interpretation, advanced by Charles S. Peirce, that probability refers to propositions about events and not to the events themselves. Thus, instead of saying that an event is likely to occur p times out of q we say instead that, if we make the proposition that the event will occur, then we shall be right p times. This overcomes certain difficulties of interpretation since it refers to our expectations based upon such knowledge as we possess and does not accord to the concept of chance an identity of being some separate positive force which employs or obeys rules of its own. The theory of probability is a wide one and calls for some very clear thinking before its rules may be applied in statistical inference. It is a subject in which it is essential to think back to basic principles.

The probability of a specified outcome of specified events is equivalent to the relative frequency of that particular outcome observed in all events of exactly the same kind. It obviously helps enormously if we know in how many ways an event *can* occur before presuming to predict how it *might* occur. Many probabilities are self-evident in this way since the total number of possible outcomes is fixed. A dice has six faces and, provided the dice is symmetrical and not loaded, each of these faces has an equal chance of appearing uppermost when the dice comes to rest. There are six – and only six – possible results; the dice cannot for example stand on one of its edges or corners. Since there are only six results possible and since each face has an equal chance of appearing uppermost, then the probability that a specified face will appear is one in six.

This probability may be measured because the total number of possible results is known. Many statistical problems, however, are concerned with circumstances in which the total number of possible results is not known. If the number of possible outcomes of rolling the dice had not been known, might one have argued that the probability of any specified outcome was one in two? Either a specified face would appear or it would not; there would be two possibilities only. From the previous calculation we know that this is a fallacy, but why is it not true? The fallacy arises in the assumption that the occurrence and the non-occurrence of an event are necessarily equiprobable. There may be one way in which a desired result may be achieved and, as with the dice, many more ways in which results other than the desired result may be achieved. If the thrower is trying to throw a six, there is only one way in which this may occur whereas there are five other ways in which it may be prevented. The occurrence and the non-occurrence are not therefore equiprobable at all.

Some apparently self-evident probabilities are therefore fallacious. If you toss a penny you would reasonably expect that the probability of a head falling uppermost would be one in two. There are only two possible results – heads or tails. Just as a dice will not stand upon one edge, neither will a penny do so. Can the same probability ratio be ascribed to the proposition that an expected baby will be a boy? A baby is either a boy or a girl; there is no other possibility for a normal child. Nevertheless this is not a self-evident probability. If the birth-rate statistics are studied it will be noted that the numbers of girls and boys born are not identical. In 1958 there were 1,058 males born in the United Kingdom for every 1,000 females. Records back to 1870 show that in every year there have been more male than female births, the ratio varying between 1,035 and 1,065 males for every 1,000 females.[1]

The probabilities attending the sex of newborn children is therefore not self-evident, although they are often said to be so. The difference between the ratios may be comparatively slight, but there is a difference. An Australian doctor of science is reported[2] as saying that the overall world ratio between births was 105·5 boys to 100 girls 'whereas according to science, it should be a 50–50 chance'. Science says no such thing. Just because a child is either a boy or a girl it does not necessarily follow that the births are equiprobable – the actual figures show that they are not. The doctor is also reported as saying that nature was bringing more boys than girls into the world to maintain balance between the sexes since men tended to live shorter lives than women. He thus admits that there is a difference in the respective birth-rates and appears to suggest that nature is achieving this in defiance of science. But science cannot be at variance with measurements of known natural phenomena, and the scientist who thinks it is at variance needs to do some urgent re-thinking.

A penny may behave in one of two ways, each of which is equiprobable. A normal baby does not enjoy the same 'choice' of action. A penny is free to react in different ways right up to the action of being tossed. A baby's sex may not be known until it is born, but it is predetermined before that stage. The event has already occurred and we are ignorant of it only because we have not been admitted to the secret. This again emphasizes that the probability calculations refer to our propositions as to a baby's sex rather than to the sex determination.

In calculating the probabilities involved in the tossing of a penny we were always considering the same penny. We can, however, apply the results to each and every penny which is identical with the original one. A deformed penny might tend to fall more often upon one particular side but normal unloaded pennies may reasonably be expected to react in identical ways. Pennies share a common identity. Babies, however, although

1. *Annual Abstract of Statistics* (1959), H.M.S.O.

2. *Evening Standard*, 16 November 1960.

possessing certain obviously common human characteristics, are not identical even though they may be indistinguishable except to their proud parents. Each child is the offspring of its parents and the family histories of the latter may show a propensity for a particular sex to predominate. In such circumstances it would not be correct for probabilities calculated for a group to be applied to a specific case since to do so is to particularize from the general. It may be said that to take a general probability for the tossing of pennies (irrespective of which pennies are tossed) and to apply it to a particular toss is to follow the same invalid inference. It is certainly drawing a particular inference from a group probability but it works because every penny is apparently identical.

It is conceivable that a coin might be tossed in such a way that by practice an individual could obtain a desired result more often than not. In such circumstances the probabilities involved are not the same as those involved in non-controlled tosses for the excellent reason that the controlled tosses and the non-controlled tosses form quite distinct statistical sets. An analogy may be drawn here. There are about 27 million females in England and one Queen. This does not mean that every woman has one chance in 27 million of becoming Queen. All of them, except for those in the Royal Family, have no chance whatever of becoming Queen for the simple reason that they are not members of the same group.

If the problem of the pennies could not have been solved by self-evident or *a priori* methods, it could have been approached in quite a different way by experimenting. Toss one penny and note the result. Repeat this a dozen or so times. The first three tosses may all reveal tails and doubts may spring up as to the accuracy of the one-in-two probability. But mathematical probability and reality have no common ground in respect of very small numbers. As the number of tosses is increased, however, so will the actual results approximate more closely to the one-in-two pattern. Thus in 2,000 tosses there may be 998 heads, whereas in 20 tosses there may be only 8 heads. The larger the number of experiments so the more closely will the true probability be disclosed. The probabilities calculated refer to the general behaviour of the members of a statistical set. The probability of a head being tossed is one in two but it does not follow that, in order to maintain the probability ratio, the next toss will produce a tail. Similarly, although the numbers of boys and girls born in the United Kingdom are very nearly equal, it does not follow in the short period in a small community that the numbers of boys and girls born will bear even the same approximate ratio to each other.

The empirical or experimental method may also be applied to events about which we can have no *a priori* ideas. Some values do not belong to finite statistical populations. Records over a period will reveal on how many occasions a wind gust of 85 m.p.h. was experienced along the east coast of Britain, but there is no finite total of possible occasions upon which the gusts might have been experienced. With a finite population it is possible to discover how many times an event occurred and how many times it did not occur. Where there is no finite limit it is not possible to do this. We cannot, for instance, perform a calculation like this:

Number of gusts	20
Number of non-gusts	40
Probability of gust	One-in-three

Non-gusts cannot be measured, for the very reason that they never existed and we have had to resort to inventing a name to describe them. Nevertheless the occurrence of the gusts is very real and it is desirable to be able to estimate the probability of such occurrences related to accompanying weather or seasonal conditions.

The empirical method is also useful if, although we are dealing with a finite population, we do not know the size of the population. If there is one red ball and eleven black balls in a bag and if each ball has an equivalent chance of being selected, then the probability that the red ball will be selected is one in twelve. There is only one way in which the red ball may be selected but there are eleven ways in which its selection may be prevented. This is self-evident. But, if we do not know how many black balls there are in the bag, the position is then not self-evident since we do not know in how many different ways the selection of the red ball may be prevented. By experiment, however, we may find that for every time we draw a red ball there are about eleven times when we do not. This would tell us that the probability is one in twelve of selecting the red ball, provided the number of experiments is sufficiently large, and that this probability may be applied to future calculations. This, incidentally, gives us a pretty good idea about the number of balls in the bag, but we can manage quite well without knowing the exact figure.

Probability calculations are based upon hypotheses which may thus be tested by experiments, but it is important to remember that this does not give a logical proof of the accuracy of the ratios assigned. Experiments merely show whether or not the relative frequency approaches the predicted probability, but they also provide the basis for further predictions. It is this aspect of the matter which is so valuable because in statistical research it is often the case that the total number of possible outcomes of an event is not known.

In mathematical terms the probability of an event is measured as having a value between 0 and 1. An event which is certain is assigned the probability value of 1, whereas an event which is impossible has the probability value of 0. Strictly speaking, therefore, neither 0 nor 1 measures probability in the non-statistical sense. Probability in the general sense involves doubt, and anything that is certain has no degree of probability. The mathematician, however, uses these two values to limit the values for all probabilities. The probability that a head will lie uppermost after a penny has been tossed is expressed mathematically as the ratio between the total number of ways in which the desired result can be achieved and the total number of all possible results.

$$p = \tfrac{1}{2} = 0{\cdot}5$$

While we cannot be certain whether a head or a tail will appear in a particular experiment, we are certain that it will be either a head or a tail. The probability of a head is 0·5, as also is the probability of a tail. The total probability of all outcomes is thus

$$p = 0{\cdot}5 + 0{\cdot}5 = 1$$

This additive process derives from the addition law for mutually exclusive events. The penny cannot show both a head and a tail at the same time. The two possible outcomes of the event cannot occur together; if one occurs, then the other cannot occur and they are therefore mutually exclusive. If a bag contains 3 red, 5 black, and 2 white balls, then the respective probabilities are:

picking 1 red ball	$p = \frac{3}{10} = 0{\cdot}3$
„ 1 black „	$p = \frac{5}{10} = 0{\cdot}5$
„ 1 white „	$p = \frac{2}{10} = 0{\cdot}2$
picking 1 ball of any colour	$p = \frac{10}{10} = 1{\cdot}0$

similarly, by the addition law, it is possible to calculate the probability of success if, say, it is sufficient to pick either a red or a black ball. This probability is the sum of the individual probabilities shown above

$$p = 0{\cdot}3 + 0{\cdot}5 = 0{\cdot}8$$

This addition is possible because the individual outcomes are mutually exclusive. If they were not so exclusive they could not be combined by addition. If one is forecasting the result of a horse race in which eight horses are running, and if it is assumed that each horse has an equal chance of winning (this is rarely true in real life!) then one of the eight horses must win to the exclusion of all the others, and the forecaster will have one chance in eight of selecting the winner. The probability that he will do so is therefore $\frac{1}{8}$ or 0·125.

Now repeat the same set of circumstances for the second race. Again the probability of success in that race will be 0·125. What, then, is the probability that he will forecast at least one winner from the two races? He had one chance in the first race and he has the same in the second race. He has apparently doubled his chances; has he at the same time doubled the probability of success? In fact he has not. In order to demonstrate this point, let it be assumed that the forecaster attempts to select the winning horses in ten races, in each of which there are eight horses running. If the forecaster increases the probability of success by increasing the number of races, then the total probability of success would be the total of all the separate probabilities for the individual races. There are ten races in each of which the probability of success is 0·125. The total probability will therefore be 1·25, but this is impossible since no probability value can exceed 1 which represents certainty!

Another way of demonstrating this point is to show that, if a forecaster's probability of success increases progressively as the number of races increases, then he only has to forecast the results of eight races and he will be certain to pick a winner. The addition of the separate probabilities from eight races is equivalent to $8 \times 0{\cdot}125$ and this is equal to 1 or absolute certainty. There is evidently something wrong here. The truth is that although the forecaster might appear to be increasing his chances of success he is also increasing his chances of failure and this leaves him exactly where he was. If he has one in eight chances of success in each race this merely means that on average, over a sufficiently large number of races, the number of his actual successes will be approximately equivalent to one-seventh of the total number of actual failures. He will not necessarily enjoy one success in his first eight races, but even if he does so he cannot be said to be increasing his chances of success with each additional race – the success could have occurred in any one of the eight races.

There is, nevertheless, the nagging question of deciding, when there has been a run of results apparently contrary to probability expectations, whether the probability of a change in the results does not approach more closely to certainty as the run grows longer. This, however, is approaching the deep waters of considering the probability of a probability. The short answer, of course, is that the result of one race or a series of races can have no effect upon the result of a subsequent race since the races are entirely independent of each other. This is why the addition law cannot be used; the results are not mutually exclusive. A man may successfully forecast the winner in each of the races. He is not precluded from forecasting the winner of the second race merely because he has already forecast the winner of the first race.

A simple form of the same kind of problem is posed by the question – what is the probability that a head will appear at least once in two tosses of a coin? The French mathematician D'Alembert[1] argued that there were only three possibilities – a

1. See Todhunter's *History of the Theory of Probability*, 1865.

head would appear on the first throw, or on the second throw, or not at all. Two of these possibilities would give the desired result and the probability of this was therefore $\frac{2}{3}$. This argument, however, ignored the fact that there were four possible results, not three as supposed by D'Alembert.

	1st toss	*2nd toss*
Possibility 1	Head	Head
2	Head	Tail
3	Tail	Head
4	Tail	Tail

A head or a tail on the first toss may each be combined with either a head or a tail on the second toss. There are four possibilities; in each of three of these there is at least one head and the probability of throwing at least one head is therefore $\frac{3}{4}$. It has been necessary to go back to the basic data to calculate how many different possible and equiprobable results may be obtained and how many of these will satisfy the required result.

The combination of probabilities of independent events is achieved by the multiplication law. Here we are not concerned with the total probabilities of all events within a class of a limited number of possibilities. Instead we are concerned with the combining of separate probabilities for individual events in different classes. Separate races are separate classes. If the probability of selecting a winning horse is $\frac{1}{8}$ in each of two races, then the probability of selecting both winners is

$$\frac{1}{8} \times \frac{1}{8} = \frac{1}{64}$$

The reason for this is that every one of the eight horses in the first race may be paired off with each of the eight horses in the second race. There are therefore 64 possible pairings, only one of which will contain both winners. The multiplication law thus states that the probability that two independent events will both occur is the product of the individual event probabilities.

Two events may also be independent of each other even if they form dependent stages of one sequence of events, provided they are clearly identifiable and that, once one of the events has occurred and has perhaps caused a change of circumstances, it can have no further effect beyond this on the next event. The first event may restrict the functioning of the second event. A man may buy some cigarettes and sweets at different shops. In buying the cigarettes he has reduced his stock of money and has restricted the limits of expenditure within which he can purchase the sweets, but the two purchases are nevertheless quite separate events. The same principle applies in probability theory where the two events are described as being conditionally independent. In a bag there are six black balls and three blue balls. If we select two balls at random, what is the probability that we will take one black and one blue ball?

The first ball will be either black or blue. The probability that it will be black is $\frac{6}{9}$, since there are six ways of selecting a black ball out of the total nine. If a black ball is thus selected first, then we shall be left with a changed 'bagful' of five black balls and three blue ones. The probability of then selecting one blue ball is thus $\frac{3}{8}$ since there are three chances out of a total of eight. Consequently the probability that we shall pick one of each colour is calculated by multiplying the two separate probabilities

$$\frac{6}{9} \times \frac{3}{8} = \frac{1}{4}$$

To check this calculation, try it the other way round. The probability that we shall select a blue first is $\frac{3}{9}$ and that we shall then select a black is $\frac{6}{8}$. The combined probability of selecting one of each is then

$$\frac{3}{9} \times \frac{6}{8} = \frac{1}{4}$$

The applications of these two laws – addition and multiplication – are quite distinct. They give different types of answers because the questions are different, and it follows that, if one wishes to find the correct answer, then one must know what question to ask. Before probability rules are applied to problems it is first necessary to decide that they are in fact appropriate. An obviously incorrect result will show itself up, but other errors might remain undetected. The addition law, for example, may be applied directly only to mutually exclusive and dependent events.

There are three women waiting to be interviewed for a clerical post, their abilities being as follows:

Mrs A.	Shorthand	Typing
Miss B.	Typing	Clerical
Miss C.	Filing	Shorthand

What is the probability that the first one to be interviewed is proficient at either shorthand or typing?

Two out of the three are proficient at shorthand so that the probability of the first interviewee being in this classification is $\frac{2}{3}$. Similarly two out of three are proficient at typing, so that the probability here is also $\frac{2}{3}$. If we add these two probabilities together, we obtain a result that the combined probability is $\frac{4}{3}$, which is manifestly absurd. The reason is that the events are not mutually exclusive. Mrs A, for instance, is proficient at both shorthand and typing and she has therefore been included in the calculations twice. The true probability is the sum of the individual probabilities less the probability that the two proficiencies will occur together in one individual. The probability that they will occur together is $\frac{1}{3}$, so that the combined probability that the first interviewee will be proficient either at shorthand or typing is

$$\frac{2}{3} + \frac{2}{3} - \frac{1}{3} = 1$$

This is equivalent to certainty and may be confirmed by checking back to the list of proficiencies; every candidate can either type or read shorthand.

In probability calculations it is essential that all possible outcomes which appear to be equiprobable should be proved to enjoy that equality. We have already noted that the probability of Mrs Bloggs becoming Queen is not the same as the probability enjoyed by a princess of royal birth. Many statistical problems, however, are not so clearly centred upon outcomes which are not equiprobable.

Of three bags, one contains two red balls, one contains two black balls, and the third contains one ball of each colour. Then, if the bags are indistinguishable according to their contents, the probability of selecting the bag containing the two unlike balls is $\frac{1}{3}$, since there are only three possible choices. Having chosen one of the bags, we take out one of the balls and put it out of sight without looking at it. Then, irrespective of which bag we chose and which ball was taken out, the remaining ball in the bag is either black or red. Further, it must be either like or unlike the ball which was taken out. The probability of their being unlike is $\frac{1}{2}$ and therefore the probability of selecting the bag with the unlike balls must be $\frac{1}{2}$. We have already seen, however, that the true value for the latter probability is $\frac{1}{3}$. Where has the calculation erred? This paradox was considered by the French mathematician J. Bertrand[1] who pointed out that the fallacy lay in assuming that when one ball had been taken out of the bag the possibilities then resulting were equally likely. He showed that if the ball taken out of the bag is black, the second ball is less likely to be red than black. How does this follow?

Bag	*Contents*	
1	Red	Red
2	Black	Black
3	Red	Black

If the first ball taken out is black, it cannot have come from bag 1. It must therefore have come from either of bags 2 or 3. The probability that the first ball taken from bag 2 is black is equal to 1 – the event is certain. On the other hand, the probability that the first ball taken from bag 3 is black is equal to $\frac{1}{2}$ since there are only two possibilities. Therefore, if a black ball is drawn first it is less likely to have come from bag 3 than it is to have come from bag 2. Consequently the second ball is less likely to be red (that is, the other ball in bag 3) than it is to be black (the other ball in bag 2). Similarly, if a red ball had been taken out first, it is less likely that the second ball would be black. The events are therefore not equiprobable and the probability ratios should not be calculated upon the assumption that they are.

Reference has already been made to the fact that a run of results does not affect the individual results of independent events.[2]

1. *Calcul des probabilités*, 1889.
2. See page 198.

Mathematical probability can be applied to reality only in long runs of typical events. Although the probability of tossing a head is one in two, it does not follow that every other throw in a consecutive order of throws will produce a head. The same principle applies to the calculation of combined probabilities. The probabilities of throwing 1, 2, or 3 heads consecutively are:

1 head	$p = \frac{1}{2}$
2 „	$p = \frac{1}{2} \times \frac{1}{2} = \frac{1}{4}$
3 „	$p = \frac{1}{2} \times \frac{1}{2} \times \frac{1}{2} = \frac{1}{8}$

When we have thrown two heads in succession, what is the probability that the next throw will also be a head? The probability of throwing three consecutive heads is $\frac{1}{8}$. The third throw is part of this run of three throws. The probability of $\frac{1}{8}$ applies to the outcome of the whole run and thus to the outcome of the final throw. But consider the third throw entirely on its own merits. This is merely a repetition of an event in which the probability of a head is $\frac{1}{2}$. Thus when we have thrown two heads, since the probability of throwing a head on the third throw is $\frac{1}{2}$, then the probability of completing the set of three consecutive heads is also $\frac{1}{2}$. The true probability that three heads will be thrown, however, is $\frac{1}{8}$. How does this discrepancy arise?

The answer is that these two probability ratios measure different things. The third throw must be considered on its own merit alone and not as part of a run, since the probability of any outcome of the third throw cannot be affected by preceding events. Actual occurrences in the past cannot be added to the probability of future occurrences so as to give a total probability for the set of all the occurrences. Probability has no meaning for the past. The events have already occurred. They rank as certainties and no question of how probable they are can possibly arise.

But does this not make nonsense of the original statement that the chances of three consecutive heads are one in eight? How can the probability of an event change half-way through its course? If the original probability is one in eight, why is it that when we have already thrown two heads, the probability increases to one in two? Probability theory is concerned with chance or with the unknown factors which produce the results. Once events have taken place and the outcomes are known, the mathematical probability in each case is 1. The probabilities are therefore:

Head on first throw	$p = 1$
,, ,, second ,,	$p = 1$
,, ,, third ,,	$p = \frac{1}{2}$

and the probability of completing a run of three consecutive heads is therefore equal to $1 \times 1 \times \frac{1}{2} = \frac{1}{2}$.

Having actually thrown the first two heads, we have eliminated the adverse chance factors in those two throws. The probability of complete success improves with each successful stage completed just as a tennis player's chances of winning a tournament improve with each match won, assuming unkindly that players are so well matched that the results depend upon chance.

The multiplication law can give rise to results which at first sight are almost unbelievable. Suppose that there are thirty people at a meeting. What is the probability that two of these will have the same birthday (that is, date of day and month but not necessarily of year)? There are 365 days in an ordinary year and there are therefore 365 different possible birth-dates. What would you expect the chances of a duplication to be? One in twelve perhaps or, say, one in ten?

The easiest way to calculate this ratio is by first calculating the probability of total failure; that is, the probability that there is not a duplication of birth-rates. The first man has 365 'possible' dates. The second man therefore has 364 chances out of 365 of not duplicating the first man's date. Similarly, the third man has 363 chances of not duplicating either of the dates of the first two men. The chances of failure on the part of the remaining twenty-seven men reduce by 1 in 365 at each remove so that the thirtieth man has 336 chances in 365 of not duplicating any of the dates of the preceding twenty-nine men. The total probability of failure therefore is the product of the twenty-nine terms:

$$\frac{364}{365} \times \frac{363}{365} \times \frac{362}{365} \times \ldots \times \frac{336}{365}$$

and this is approximately equivalent to 0·3. This, however, is the probability of failure. The probability of success or failure is 1, since one of the results must be achieved, and the probability of success is therefore

$$1 - 0{\cdot}3 = 0{\cdot}7$$

There are therefore approximately seven chances out of ten that there will be a birth-date duplication. This is quite out of character with what might have been considered probable without the assistance of mathematics and it serves to demonstrate that probabilities should never be assessed intuitively unless the intuition is really a manifestation of experience.

In all the foregoing problems we have assigned finite limits to the calculations by stating the number of events to be considered. That is, what results may be obtained in three tosses of a coin or in two selections of coloured balls or amongst thirty people. All these are finite classes. A famous problem – the St Petersburg Paradox – will illustrate the kind of difficulty which may be encountered where there are no finite limits. We return to our coin-tossing with the assistance of Mr A and a Banker. Mr A tosses a coin and, if a head appears, the Banker pays him £1 and the experiment is over. If, however, Mr A does not get a head on the first throw he continues to throw until he does. With every throw his possible prize is doubled. Thus if he tosses a head on the second throw, he will win £2; on the third throw he will win £4 and so on. The problem is to assess what amount Mr A should pay the Banker for the privilege of playing, so that the game shall be a fair one, neither Mr A nor the Banker having an unfair advantage no matter how long the game continues.

The probability of a head on the first toss is $\frac{1}{2}$. The prize at this level is £1 and the value of Mr A's expectation is therefore 10 shillings. Mr A will win on the second toss only if his first throw was a tail and his second throw was a head, and the probability of this combination of results is $\frac{1}{2} \times \frac{1}{2}$. The prize at this level is £2 and the value of Mr A's expectation is therefore £2 × $\frac{1}{4}$ which is again equal to 10 shillings. Similarly the value of his expectation is 10 shillings at every toss and he must therefore pay ten shillings for every toss. Theoretically there is no limit to the number of tails that may be thrown consecutively. At, say, the thirtieth toss, the expectation will have risen to £15 (that is, thirty times ten shillings) and, as there is no theoretical limit to the number of tosses, then the overall total expectation is also infinite. The player must therefore pay to the Banker an infinite amount of money for the privilege of playing one game.

The game of double or quits has gone mad! The mathematical working is correct yet the result is patently absurd. There are a number of objections, all of which point to the fallacy of allowing a strictly correct mathematical result to get the better of common sense. We have been discussing a theoretical infinity, but the problem posed is a practical one. It involves the actual tossing of a coin and the possibilities involved, although not calculable, are nevertheless finite. The first objection is that in practice a coin will not continue infinitely to show a tail in every toss. Each run of tails will in fact be limited at some level. A second objection is that a game of infinite length would outlive both Mr A and the Banker or they would have to live eternally to continue the game.

Again, although the requirements of an infinite game would involve Mr A in the expenditure of an infinite amount of money, the amount of prize money payable by the Banker also increases infinitely. In fact it will increase rapidly to astronomical proportions. If a head is not thrown in the first thirty throws, the prize money payable will have increased to $(£2)^{29}$ or more than £536 million. Imagine what the prize money will have increased to after a hundred throws. The Banker's stock of money at the commencement of the game would be limited, and to hold out prospects of a prize greater than that amount – that is, an infinitely large prize amount – does not conform to the moral requirements of fairness. The Banker could not possibly pay a prize greater than the money in his possession, and he cannot therefore ask Mr A to pay in expectation of winning a prize which the Banker could not provide.

A compromise suggestion is that any probability lower than, say, one in ten thousand should be treated as if it were zero. This would limit the game to a finite number of tosses. The probability of tails appearing for n successive tosses is equal to $(\frac{1}{2})^n$. As soon, therefore, as n becomes so large that $(\frac{1}{2})^n$ becomes less than $\frac{1}{10{,}000}$, the player's expectations would cease to increase. This would occur at the fourteenth throw and Mr A would be expected to pay £7 to enter the game (that is, 14 throws at an expectation of 10 shillings per throw). Upon Mr A's having lasted the course, the Banker should admit that a head must come down some time and should pay out the prize money already accumulated. This amount, incidentally, would then be $(£2)^{13}$ or £8,192. Tossing experiments have been carried out *in extenso* and in a sequence of over 8,000 tosses, two runs each of fifteen consecutive tails were observed, so that an arbitrary limit of fourteen tosses in the game would not be unreasonable.

The finer points of this problem are perhaps more for the number theorist in pure mathematics than for the statistician, but it has at least one important moral for the latter. The mathematics show what *might* have happened in unreal conditions, but not what could happen in real circumstances. The statistician is certainly required to calculate what might happen in given circumstances, but he should relate his thoughts to the possibilities involved in the practical applications of his theorizing. It may be necessary to assign arbitrary limits to an apparently infinite series.

Mathematical probability ratios represent a form of perfection which is rarely if ever achieved. The basic mathematical properties of probability are generally accepted as providing useful working bases for calculations, but the mathematics then have to be interpreted and it would be unrealistic not to admit that this is the sphere of philosophy providing scope for a considerable amount of controversy. Nevertheless, results in the forms desired can often be produced and acted upon provided it is remembered that they record probabilities and not expected actual results in isolated instances. Subject to this qualification, a probability backed by executive action may become a near certainty.

15 Normal and Other Distributions

Statistics is able to infer a great deal of useful information about populations – human or otherwise – largely because of the tendency to conformity in the grouped reactions of those populations. Individual reactions will obviously not be identical, but the frequency of the occurrences of all reactions studied together does tend to conformity. It is this fact that enables us to apply probability theory to statistical data. Clearly if conformity were entirely absent it would not be possible to make any inferences at all.

Although the measurements of a variable will show some differences, many of the values will be repeated over a long period. In the processing of data it is therefore helpful to collect the identical values together and to show the frequency with which each individual value is observed in the total number of observations. The schedule of all the frequencies of all the values observed is called a frequency distribution, this merely meaning that the schedule reveals how the frequencies of the values are spread or distributed over the full range of values. Where there are very many observations or where the difference between values is slight, it is usual to condense the tabulation into groups or classes of frequencies. Thus, while an ordinary distribution would show separate frequencies for each of the values 1, 2, 3, . . . 10, a grouped distribution would show total frequencies for, say, each of the two groups 1 to 5 and 6 to 10 respectively.

Here it is desirable to distinguish between discrete and continuous variables. A continuous variable can take any mathematical value, including a fractional one, within the range of the distribution. A discrete variable, however, can only take values which differ among themselves by certain fixed amounts. Temperature, for example, is measured in degrees, but changes between the measurements in fact pass through a continuous scale even though the changes may be so small as to defy practical measurement. Temperature is therefore a continuous variable, whereas the counting of the number of matches in match-boxes will give measurements of a discrete variable. The counting jumps in stages of 1, 2, 3, and so on. There cannot, for instance, be 49·3 matches in a box – there may be either 49 or 50 – but, allowing for imperfections of the measuring equipment, it is theoretically possible to record a temperature of 49·3 degrees.

Thus, for the discrete variable of numbers of matches in the numerical range 49 to 50 there are only two possible measurements – 49 and 50 respectively. For the continuous variable of temperature, there is an infinity of 'possible' measurements between 49 and 50 degrees although, for practical purposes, it is true that they cannot all be measured. The number of practicable measurements of the variable is restricted by the limitations of the measuring process; during the rise of temperature from 49 to 50 degrees, the temperature will at some point actually be 49·0001 degrees but this could not be measured as such.

However, since continuous variables can theoretically have infinitely small differences between the observed values, it is usual to classify these into grouped frequencies in order to facilitate calculations. Discrete variables do not always need to be grouped into classes, since they themselves are more clearly definable, but where the range of values is wide it is often convenient to do so. There is thus a basic difference between discrete and continuous variables but for many purposes this difference may be ignored. When the values of a continuous variable are grouped into frequency classes, their distribution becomes a frequency distribution. Frequencies may only be measured in terms of whole numbers since it is impossible, for instance, for the value x to occur, say, one and a quarter times. All frequency distributions are therefore discrete, whatever the character of the underlying variable. Similarly, the limitations of measuring ability, to which reference has already been made above, inevitably result in the actual measurements recorded being discrete even when the variable measured is actually continuous.

On the other hand, discrete variables may often be treated as if they are in fact continuous. One of the elementary differences between the types of variable lies in the fact that the values of a continuous variable theoretically merge imperceptibly into each other and can therefore be represented graphically as a smooth curve. A discrete variable, however, changes value in jumps and can be represented exactly only in the form of a histogram where the steps between values are obvious to the eye. Histograms are very useful as pictorial representations of absolute facts, but they do not conform to mathematical model requirements in the same way as smooth curves when it is desirable to relate these facts either between themselves or to other variables.

Because of this, the statistician calls into use the concept of geometrical approximation. The area of a square inscribed within a circle will not approximate closely to the area of the circle, but the area of a five-sided polygon will be seen to approximate more closely to the area of the circle. As polygons with greater numbers of sides are inscribed within the circle, it will be seen that their areas progressively approach nearer and nearer to the area of the circle. The process may theoretically be continued indefinitely. No polygon will ever coincide exactly with the circumference of a circle since by definition a polygon has straight sides and therefore cannot be circular. Nevertheless one can conceive of a polygon having one thousand sides. It would probably not be possible to draw such a polygon, but if it could be drawn it would be practically indistinguishable from the circle; so indistinguishable in fact that for practical purposes it could be treated as if it were a circle.

Much the same sort of principle is used in justifying the treatment of discrete variables as if they are continuous. If histogram-bars are drawn progressively narrower, the tops of the bars will eventually become indistinguishable from mere points and the area contained within the curve drawn through these points will approximate closely to the area of the histogram. Where the width of the bars is narrow, the approximate curve is easily drawn but great care is necessary in drawing the curve if the bars of the histogram are very wide. Nevertheless, this theory enables the statistician to overcome the technical difficulties presented by discrete variables, and the treatment of such variables as if they are continuous will very often give results which approximate sufficiently closely to results obtained by much more complex methods.

The most common distribution is that known as the normal distribution which may be represented graphically as a symmetrical bell-like curve (as in figure 35). Actual distributions are rarely if ever exactly symmetrical, so that the normal curve represents an idealized form of distribution, but many distributions conform very closely to the normal so that they may be treated as possessing normal characteristics.

In statistics, the word *normal* has a very restricted meaning. In different spheres of activity, normality may be defined in different ways and may indeed have a totally different meaning according to the circumstances in which the word is used and the conditions which it seeks to describe. In common usage, the word *normal* may be synonymous with *ideal* or *natural* and a person is considered normal for as long as he acts in conformity with accepted standards of how he is conventionally expected to act. He is adjudged normal only if he behaves himself as he ought to do within definable though not necessarily strict limits. The statistician, however, is not a moralist nor primarily a psychologist. He is not concerned so much with *how* people ought to act as with how in fact they *do* act. The connexion between all definitions of normality is that the reactions of a great majority tend to be accepted as normal in contrast to those of a particular minority. Similarly statistical normality is recognized by the conformity of a distribution to a commonly occurring particular type of distribution.

The hump-backed nature of the normal curve emphasizes that the majority of values cluster around the centre measure and, in a truly symmetrical unimodal distribution, the mean, mode, and median will all coincide. None of the latter, however, should be regarded as the 'normal' value since it is doubtful whether any single value may be said to mark normality. Instead, the central values mark the pivot around which the range of values cluster in a normal fashion. Statistical normality is thus an attribute of a distribution and not of its individual component values. It provides a pattern to which many distributions may be compared, not because they ought to conform but merely because, in fact, so many of them do conform. Finally it should perhaps be noted that there is nothing necessarily abnormal about distributions which do not conform to the normal curve; they may merely belong to a different 'family' of distributions.

The normal distribution was first derived by de Moivre as the limiting form of the binomial distribution[1] and was subsequently rediscovered by Laplace and Gauss and became known as the Gaussian or Normal law of errors in connexion with the distribution of accidental errors in astronomical and other scientific measurements. Some scientific data closely approximate to the normal but not all of them do so. The normal law assumes that the deviations of the values from the mean value result from a large number of minor deviations, that positive and negative deviations are equiprobable, and that the most probable value of all equally good measurements is their arithmetic mean. For many years this law was accepted as having an almost universal application but this is no longer so. It is, however, still used to a great extent, unless there is some evidence that its use is inappropriate, and it has been said that 'the role of the normal distribution in statistics is not unlike that of the straight line in geometry'.[2] So useful is it in fact that, even if a particular distribution is not normal, some transformation of the data may be attempted in order that a normal representation may be achieved. This is, for example, sometimes made possible by charting the logarithms of values instead of the values themselves. Furthermore, even although a population may not be normal, the distribution of the means of samples of a given size drawn from the population usually approximates more closely to a normal distribution.

There is not just one normal curve nor is it appropriate to refer to *the* 'normal curve' in general terms as if it were a specific curve. The description refers to a type of curve rather than to each individual curve itself. There is however a standard normal curve for which many mathematical properties have already been calculated and tabulated. Other normal curves may be related to

1. See page 218.

2. A. C. Aitken, *Statistical Mathematics*, Oliver & Boyd, 1952.

the standard so that their particular mathematical dimensions may be derived speedily from adjustments to the standard tables. The equation of the normal curve is:

$$y = \frac{1}{\sigma\sqrt{2\pi}} e^{\frac{-(x-\bar{x})^2}{2\sigma^2}}$$

where y is the height of the curve at any point relative to the

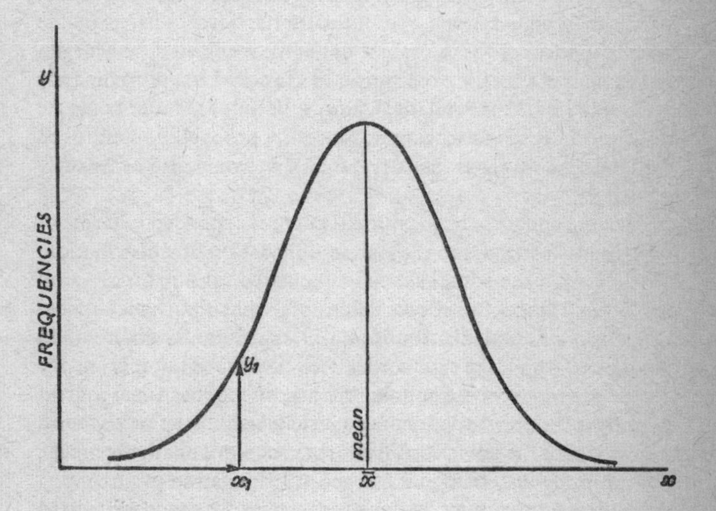

Fig. 35

The normal curve

scale of x, where σ is the standard deviation,[1] and $\bar{x}$ is the arithmetic mean of the distribution. π and e are mathematical constants familiar in many branches of mathematics. The height of the curve at any point is thus expressed as a function of the variable x together with the arithmetic mean and the standard deviation of the distribution.

Surprise is sometimes expressed that the number π, which is popularly known as a constant in the mensuration of a circle,

1. See Appendix 2.

can have other applications. De Morgan[1] once explained the chances that a certain proportion of some group of people would still be alive at the end of a given time, and he quoted a formula involving the number π. But the person to whom he was talking felt sure that de Morgan was suffering a delusion – what, he asked, could a circle have to do with the number of people alive at the end of a given time? In point of fact, of course, π is a number which can measure other things besides circle properties, just as the number seven may measure the Seven Virtues or the Seven Wonders of the World. It might have helped if de Morgan had explained that the expectation of life could be approximately represented by a mathematical curve and that the latter could be measured in terms involving π. Since the probability mentioned could be read off from the curve, then this would also necessarily involve π.

Since the equation of the normal curve is expressed in terms of the arithmetic mean and the standard deviation of a distribution, it follows that the actual shape of the curve for any distribution will depend upon those two values and that the shapes of the curves for different distributions will somehow be different. In fact they all retain the symmetrical bell-like shape but this may be elongated (where, for example, the arithmetic mean has a great majority of the recorded values clustered about it) or flattened out (as where the deviations from the mean are relatively great). It is naturally also true that the general elongation or stretching of a normal curve may be achieved merely by exaggerating the scale of the drawing, but it is only the appearance of the curve which is altered; its basic properties remain unchanged.

The area below the curve can also be calculated, so that it is possible also to calculate the total density of frequencies contained between any two y ordinates. For instance about 68% of the total area is contained within ordinates drawn on each side of the mean at 1 standard deviation distance from the mean. Thus, about 68% of the total number of frequencies lie within the same ordinates. About 95% lie within 2 standard deviations on either side of the mean, and about 99·73% lie within ±3 standard

1. *Budget of Paradoxes.*

deviations. This is represented in figure 36. Other data (not shown) are that 50% of the area of the curve is contained within the limits of ±0·6745 standard deviations; and that the area between ±4 standard deviations is 99·994% of the total.

This information is extremely useful. If, for instance, the mean of a normal distribution is 100 and the standard deviation is 2, then it is known that some 68% of all observations lie between the

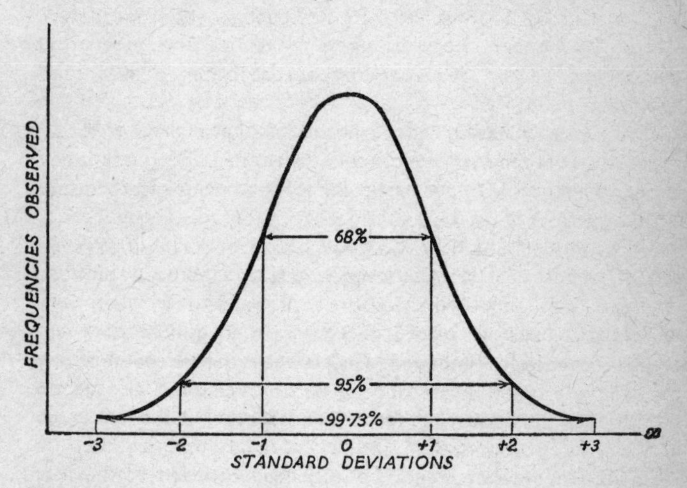

Fig. 36

Properties of the normal curve

values of 98 and 102 (that is 100 ± 2) and that nearly all the observations will lie between 94 and 106 (that is 100 ± 3 standard deviations).

The normal curve is of immense importance in sampling theory since it may be shown that the means and standard deviations calculated from random samples tend to be normal if samples are large, even if the population from which the samples are drawn is not itself normally distributed. If a number of samples are taken from a population, the value of the mean for each sample will vary within limits about the true mean of the population. The population may not be normally distributed at all, but the mean values from all the possible samples will themselves form a separate population of observations referred to as the sampling distribution. This theoretical distribution is also referred to as a probability distribution because it distributes the total probability over the different possible sample outcomes. Most of such populations tend to be normally distributed. The fact that sample statistics behave in this way gives the statistician one of his most useful tools since, once having discovered the nature of the relationship between sample statistics and population parameters, he may, if he can measure the former, thereby infer the latter.

The Gaussian law of errors postulates that chance errors in measurements tend to be normally distributed. There are many causes which can affect a particular measurement – temperature differences may affect the actual measuring device; unsteadiness of hand, however slight, may also make a contribution to error; even where the device does in fact measure correctly there may be some difficulty in reading off the measurements particularly where very small differences are recorded. These are all difficulties which cannot be entirely eliminated. The recorded measurement of any variable is a function not only of its true value but also of the efficiency of the measuring. Nevertheless, since it is the true value in which we are interested, any law which helps to explain the behaviour of measurements will also help to point to the true value. All errors cannot be eliminated, but if we can calculate an assumed true value and then, if the errors are observed to be distributed normally, there is increased confidence in the belief that the assumed value is indeed the true value or approximates very closely to it.

The important point about any formulated distribution, of course, is the fact that by its mathematical definition, wherein one variable is expressed as a function of the other, the relationship between the variables is known, whatever their respective values. If it can be shown from a sufficient number of samples that a population appears to conform approximately to a particular distribution, then the properties of the formal distribution may be ascribed to that population. It is, however, important to ensure that a population does conform to the normal since obviously only then may the population attributes be identified with the normal. Suppose that there is a machine consisting of four working parts and that a breakdown in any one of the parts also results in a breakdown in the machine itself. Then, even though the fault probability curve for each part may itself be symmetrical, the fault probability curve for the machine as a whole will not be symmetrical.

If the fault probabilities for individual parts are expressed in terms of periods of one week, so that the probability of a part fault in the first, second, and third week etc. is known, then the probability of a machine breakdown may also be calculated for the first week. The probability of a breakdown in the second week, however, is reduced by the fact that it can occur then from a specific fault only if it has not already occurred in the first week from another fault. The probabilities of breakdown in subsequent weeks are reduced still further by the probability that a breakdown will already have been caused in previous weeks. The probability of a breakdown in the first week is consequently greater than the probability of a breakdown in subsequent weeks and the probabilities reduce rapidly over the first few weeks. The distribution of these probabilities is therefore not symmetrical and it follows that any attempt to attribute normal properties to it would result in false conclusions.

The law of error was originally deduced by Laplace from the basic assumptions of the equiprobability of positive and negative deviations. Gauss later used these deductions to prove that the most probable value of any number of equally good (that is, subject only to chance deviations) observations is their arithmetic mean. This proof, however, is based upon Laplace's theories and does not prove the truth of the latter. Professor M. G. Kendall, in referring to the history of the normal distribution[1] has said that 'the discovery that errors ought, on certain plausible hypotheses, to be distributed normally led to a general belief that they *were* so distributed'. Gauss's proof does not prove the universal applicability of the law deduced by Laplace; it merely proves that if the hypotheses are correct then certain results will follow. In

1. *The Advanced Theory of Statistics.*

fact, there is no theoretical proof of the hypotheses. There cannot be such a proof because there are known error distributions which do not conform to the normal law.

The absence of formal proofs, however, does not necessarily invalidate the application of the normal law if it is apparent from the distribution itself that it does approximate to the normal. The application of the law is justified in that 'in representing many types of observations it is apparently not far wrong, and is much more convenient to handle than others that might or do represent them better'.[1] This justification does not extend to circumstances in which the normal law may clearly be shown to be non-representative of a distribution, but if the bell fits it is unwise to ignore the message of its chimes.

THE BINOMIAL DISTRIBUTION

The binomial distribution is symmetrical in its limiting form, but not otherwise. This distribution is so called because of its relation to the expansion of the binomial

$$(p + q)^n$$

a binomial expression being one which contains two terms joined by the plus or minus sign.

The binomial distribution is a probability distribution of possible outcomes of events which may be classified either as positive or negative. That is to say, it is concerned with circumstances in which a specific event will either occur or it will not; there are no half-measures possible and no account is taken of the degree of intensity of the occurrence. The total probability of an event either occurring or not occurring is equal to 1. If, therefore, the probability of its occurrence is p, then the probability of its non-occurrence is equal to $1 - p$. Similarly if the probability of the event's non-occurrence is q then, by definition,

$$p + q = 1$$

1. H. Jeffreys, *Theory of Probability*, Oxford, Clarendon Press, 1948.

Thus, if the probability of an occurrence is one in ten, then $= 0{\cdot}1$ and $q = 0{\cdot}9$, so that

$$p + q = 0{\cdot}1 + 0{\cdot}9 = 1{\cdot}0$$

The terms p and q refer to the probability of the occurrence or non-occurrence in regard to one event. The probability that the occurrence will arise in each of two separate events is, by the multiplication law for independent events, equal to: $p \times p = p^2 = 0{\cdot}01$. In the same way, the probability of non-occurrence in each of the two events is q^2 or $0{\cdot}81$. But a new set of possibilities arises since occurrence in the first event may be associated with non-occurrence in the second event; the probability of this is equal to (pq) or $0{\cdot}09$. Similarly, the probability of occurrence in the second event associated with non-occurrence in the first event is also equal to $0{\cdot}09$. The total probabilities are therefore:

Occurrence in both events $= p^2 = 0{\cdot}01$
Non-occurrence in both events $= q^2 = 0{\cdot}81$
Occurrence in *first*;
non-occurrence in *second* $= pq = 0{\cdot}09$
Occurrence in *second*;
non-occurrence in *first* $= pq = 0{\cdot}09$

Total probability of all outcomes $= p^2 + 2pq + q^2 = 1{\cdot}0$

The total probability can thus be algebraically represented as $p^2 + 2pq + q^2$ or $(p + q)^2$. Similarly, where there are three events, the probabilities are:

Three occurrences $= p^3 = 0{\cdot}001$
Three non-occurrences $= q^3 = 0{\cdot}729$
Two occurrences and one non-occurrence $= 3p^2q = 0{\cdot}027$
One occurrence and two non-occurrences $= 3pq^2 = 0{\cdot}243$

Total probability of all outcomes $= (p + q)^3 = 1{\cdot}0$

It will be seen that each of the terms of the expansion of $(p + q)^3$ represents the respective probability of different possible combinations of outcomes from all the separate events.

Term of expansion	*Term*	*Probability represented*	
		Occurrences	*Non-occurrences*
1st	p^3	3	0
2nd	$3p^2q$	2	1
3rd	$3pq^2$	1	2
4th	q^3	0	3

Similarly, for *n* number of events, the probabilities of 0, 1, 2, 3, . . . *n* non-occurrences are represented respectively by the successive terms of the expansion of $(p + q)^n$. Alternatively, the probabilities of 0, 1, 2, 3, . . . *n* occurrences are represented by the successive terms of the expansion of $(q + p)^n$. It will be noted, in passing, that the total probability is always correctly shown as 1, since, $(p + q)$ is by definition equal to 1 and it follows that $(p + q)^n$ is always equal to 1, whatever the value of *n*.

Where it is desired to calculate the probability of *at least* one occurrence, this is achieved by adding the separate probabilities for 1, 2, 3, . . . *n* occurrences, since each of these outcomes satisfies the requirements of at least one occurrence. This knowledge should perhaps make it easier to understand a problem which was once posed to Sir Isaac Newton by Samuel Pepys. The gist of the problem was that, of three men, one essays to throw at least *one* six with six throws of a dice; the second to throw at least *two* sixes with twelve throws and the third to throw at least *three* sixes in eighteen throws. What are their relative chances of success? The unwary might be tempted to say that the probabilities of success are respectively

(a) $\frac{1}{6}$ (b) $\frac{2}{12}$ (c) $\frac{3}{18}$

and that these are all equivalent. But these probabilities are for throwing one six only; the required probability for the first man is for *at least* one six and we must therefore include all the probabilities of throwing more than one six.

The required probability is therefore equivalent to the total probability of all outcomes minus the probability of not throwing a six at all. The probability of not throwing a six on one throw is $\frac{5}{6}$; the probability of not throwing one six in six throws is $(\frac{5}{6})^6$; so that the probability of throwing at least one six in six throws is

$$1 - (\tfrac{5}{6})^6 = \frac{31031}{46656} = 0{\cdot}665$$

For the probability of throwing at least two sixes in twelve throws, we must deduct, from the total probability, the sum of the probabilities of throwing (a) no sixes and (b) just one six. The probability is therefore

$$1 - 12(\tfrac{1}{6})(\tfrac{5}{6})^{11} - (\tfrac{5}{6})^{12}$$

$$= \frac{1{,}346{,}704{,}211}{2{,}176{,}782{,}336} = 0{\cdot}619$$

The probability is thus lower than that of throwing at least one six in six throws. Similarly, the probability of throwing at least three sixes in eighteen throws is lower still.

These, then, are the mathematical answers and, of course, they are mathematically correct. But do they give the real answers to what the men's real chances of success are within the limited scope of their activities? This again poses the question of what is the probability of a probability ratio representing an actual result. Newton oddly made no reference to the fact that probability ratios are more likely to represent the outcomes of a large number of events than they are to represent the outcomes of isolated events. A man who tosses a penny has on average a one-in-two chance of tossing a head in each set of two throws. Yet he may toss a penny say ten times without throwing a head. W. S. Gilbert pointed out that 'you don't find two Mondays together',[1] but there is nothing to prevent two heads or two tails occurring consecutively. When a man has thrown 1,000 times, however, the number of heads will be close to one-half of all results. It is not, therefore, correct to suggest that the real chances are proportionately the same for two throws as they are for one thousand throws.

1. *Trial by Jury*.

The respective probabilities of success are $\frac{1}{2}$ and $\frac{500}{1000}$, each of which is equivalent to 0·5, yet the real chances in the two different numerical categories of events are not in fact identical.

The man who sets out to throw at least one six in six tries is less likely to succeed in one set of six throws than he will be in a

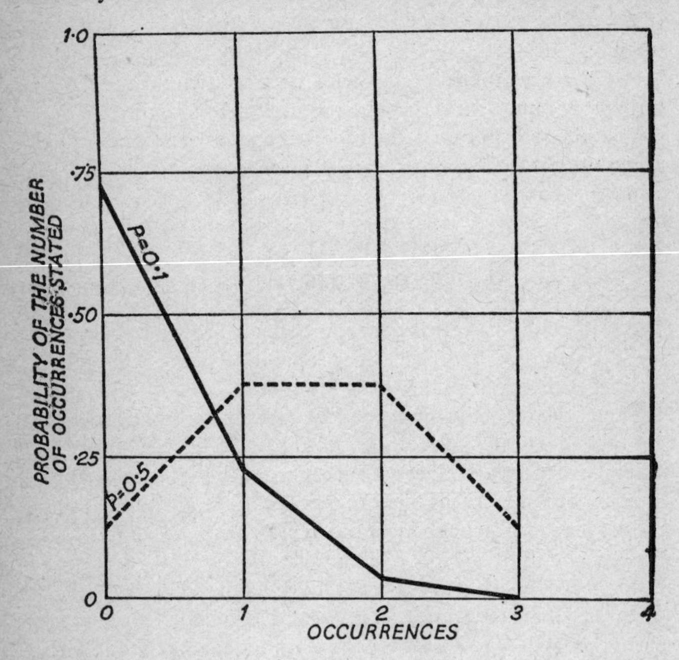

Fig. 37

Binomial distribution ($n = 3$)

succession of sets. The mathematical probability applied to events in isolation is meaningless. The mathematical probability of the second man throwing at least two sixes in twelve throws is lower than that of throwing at least one six in six throws, but the circumstances under which they are operating are so different that it is not meaningful to compare their respective chances of success.

Where $p = q = \frac{1}{2}$, the curve of the binomial distribution is symmetrical. The expansion of $(p + q)^3$ for example is

$$(p + q)^3 = p^3 + 3p^2q + 3pq^2 + q^3$$

and if p and q are equal, then so are p^3 and q^3 equal to each other, and $3p^2q$ is also equal to $3pq^2$. Where, however, p and q are not equal, the curve is not symmetrical. If, for example, $p = 0{\cdot}1$; $q = 0{\cdot}9$; the curve will be quite different. This difference is brought out by the respective frequency polygons in figure 37.

When n is very large, however, the distribution represented by the terms of the expansion $(p + q)^n$ does more closely approach symmetry.

THE POISSON DISTRIBUTION

The binomial distribution is a probability distribution referring to circumstances where there is a calculable probability of the occurrence of an event and also of the non-occurrence of the event. That is to say that it may be calculated how often an event occurs and how often it does not occur. The total possibilities are known. Not all occurrences, however, may be classified in this tidy and satisfactory manner as, for example, the occurrence of isolated events of which there is no finite total number of possibilities. Natural phenomena are the most obvious examples of this type of occurrence. It is possible to count the number of earth tremors experienced in a particular area during a specific period of time, but it is clearly not possible to count how many tremors did not occur.

In such circumstances we cannot use the binomial expansion to represent the probability distribution since, not only do we not know the value of n – the total number of possible outcomes – but, indeed, there is no finite quantity which n may represent. For occurrences of this kind the Poisson distribution is used, and it is found that, where m is the average number of occurrences of an event, then the probability of the occurrences is described by the respective terms of the expansion of

$$e^{-m}\left(1 + m + \frac{m^2}{2!} + \frac{m^3}{3!} + \ldots\right)$$

the value of e being the mathematical constant approximately equal to 2·71828. The exclamation mark (!) used in this connexion is the mathematical factorial sign meaning 'multiply the stated number by all positive non-zero integral numbers lower than the stated number'. Thus the term 4! is a shorthand way of writing $4 \times 3 \times 2 \times 1$.

The respective probabilities are thus:

0 occurrences	$p = e^{-m}$
1 "	$p = me^{-m}$
2 "	$p = \left(\frac{m^2}{2!}\right)e^{-m}$

If, therefore, the average number of tremors is equal to one per month, then $m = 1$; and the probability of occurrences per month will be as follows, calculated on the approximate value $e^{-m} = 0{\cdot}3679$.

Number of occurrences	*Probability*	*Approx. numerical equivalent*
0	e^{-m}	0·3679
1	me^{-m}	0·3679
2	$\left(\frac{m^2}{2!}\right)e^{-m}$	0·1839
3	$\left(\frac{m^3}{3!}\right)e^{-m}$	0·0613

The Poisson distribution is again, like the normal distribution, a distribution which justifies itself by its results. If there is sufficient data on which to base the calculation of the average number of occurrences within a standard period of time, then there will also be sufficient data to reveal the actual number of occurrences within each of a specific number of standard periods. A comparison of these figures for actual occurrences with those which would have been predicted by the distribution will reveal how well the probability distribution fits the distribution of the actual observed frequencies.

The Poisson distribution really relies upon the fact that the occurrences it describes are relatively rare by comparison with the number of times they might have occurred. Similarly it will be found that the binomial distribution approximates closely to the Poisson distribution where p or q is very small and n is relatively large. The Poisson distribution is in fact derived from the binomial distribution as the limiting form of the latter where the value of n tends to infinity. This is a very useful approximation since the Poisson terms are easier to calculate[1] and in many industrial and other applications it is often a fact that p is quite small relative to n.

The binomial, Poisson, and normal distributions are the main forms of distribution employed and they provide the basic tools of a great deal of statistical work. But if these are the tools, how may one judge the work which they turn out – that is, are the results derived from their use reliable? A number of safeguards or significance tests are available mainly based upon the principle of the Null Hypothesis which postulates that, as between two samples, there is no really significant difference and that such difference as occurs is attributable only to random sampling errors.

Thus, where differences occur between the values of the means of different random samples, it is desirable to know whether the samples are, in fact, drawn from the same population. The research worker for example will want to know if different results from his experiments are arising from the application of different processes or whether they merely arise from chance. The idea behind the null hypothesis is that, if the negative character of the hypothesis may be rejected, then confidence in the opposite positive hypothesis – that the observed differences are not arising from chance factors – may be strengthened. If the research worker observes a difference which could occur by chance less than once in a thousand times then he is justified in treating his results as being statistically significant – that there is a real difference because the null hypothesis is contradicted.

Similarly, if a theoretical distribution curve does represent the distribution of frequencies in the population from which a sample is drawn, then any differences between observed and theoretical

1. See R. Goodman, *Statistics*, E.U.P. (Teach Yourself Series), p. 68.

frequencies must be due to sampling errors. If, however, the differences are greater than would be expected from random sampling, then the curve of the theoretical distribution will not adequately represent the parent population from which the sample is drawn.

In these and other ways it is possible to measure the significance of results obtained. The mechanics of these significance tests are really beyond the scope of this book, but some of them are included in the Appendices. It is, however, essential to remember that the statistical significance revealed in any test must not be confused with the importance of any differences thus shown to be significant. No statistical test is, in any event, capable alone of giving complete certainty of relative importance and it is necessary that the results obtained should always be used with sound common sense. Statistics provides information for decision making but it is not necessarily able to 'prove' what those decisions should be. A difference may be real in that it is statistically significant, but there may be practical or economic reasons why decisions cannot be taken to follow the course apparently pointed out by the statistical indicator.

16 Sampling

While it is inadmissible to particularize from the general or to generalize from the particular, it is nevertheless possible in certain circumstances to generalize from a representative section of a general group or population. Within most of life's activities there is a general pattern of behaviour, albeit a changing but nevertheless recognizable one. Where such a pattern may be discerned and measured and related to a theoretical distribution, it is possible to apply mathematical probability theory to the estimating of population parameters from random sample[1] statistics.

In a particular situation an individual will react in a special way. Different individuals will experience different reactions, yet despite this it is often possible to discern some conformity in the totality of reactions of the population of all individuals. Provided that the possibility of such a relationship can be established, then the reverse procedure of estimating the reactions within the population, from a study of the reactions within a sample, will make it much simpler to ascertain desired information about the population without first having to have data concerning every individual member of the population.

The basic idea of sampling is probably almost as old as mankind and, for all we know, possibly preceded it since some lower animals appear to have the knack of tasting a proportion of the food offered to them and of rejecting the whole meal on the strength of the one portion tasted! Sampling is based on choice and selectiveness. Early man had but few possessions and he did not need to count them nor to compare one with another. But as

1. See Chapter 17.

soon as the barter system began to assume larger scale proportions, the two parties to the exchange, say, of coconuts and local barley-wine, would not have tasted every coconut and every drop of wine. Instead they would have tested a few units of each commodity and would have assumed that the rest of the units were up to the standard of the sample units which they had selected. The purchaser of the wine, for instance, could not have tested all the wine without becoming highly intoxicated and probably insensible to the whole transaction. He might, perhaps, have tasted a little from each skin of wine, but in doing this he would still have been sampling on the theory that the remainder of the wine in each skin was exactly the same as the small amounts tasted, although his judgement might have suffered as he proceeded.

It is equally impossible today to inspect every item separately. Electric light bulbs, for instance, are tested for the longevity of useful life by lighting them until the elements break. If every bulb was tested in this way, there would not be any bulbs left to put into the lampholders at home. One hundred per cent inspection like this destroys the article, and some form of sampling is necessary to avoid this wholesale destruction while at the same time producing evidence that the bulk from which the sample is drawn is being maintained at a specified level of quality. Other populations are so large that it would be physically impossible to gather data in respect of every member. Correct sampling methods therefore make it possible to gather information which would otherwise be unobtainable.

Although a particular member of a population will react to a particular situation in a manner peculiar to himself, the total number of possible different reactions is usually finite and may be very small indeed. A toothpaste manufacturer, for instance, may carry out a survey to gauge the relative shares of the market enjoyed by his competitors. The respondents will be asked what type of toothpaste they use, and the number of different replies is limited by the number of different types of toothpaste available. If a question is to be answered either affirmatively or negatively, there are only two possible answers, provided a clear-cut decision is possible at all. The fact that the number of possible reactions is limited in number in practice forces individual members of a population to conform to one of those reactions. Even where the number of possibilities is not strictly limited in theory, they do tend to be limited in fact. This basic fact is well known. Wedding presents tend to have a marked similarity and those happy couples who cannot find a use for half a dozen coffee pots or water sets will issue a list of desired gifts. They may not be able to calculate the probabilities involved, but they are well aware of the unmeasured probability that there will otherwise be some duplications.

Human beings, for all their pretensions, have a remarkable propensity for lending themselves to classification somewhere within neatly labelled categories. Even the outrageous exceptions may be classified as outrageous exceptions! The individuals within each category tend to react to identical situations in very much the same way as each other, even though there may be a wide divergence between the typical category reactions. As soon as we can classify individuals, we are on the way to being able to deal with them statistically. Since there are a number of different possible reactions, it is obviously impossible to take the reaction of one individual as being representative of an entire population; he can at most be representative of his category within the population. One member taken from each of the categories will, however, together form a group representative of all the main reactions of all the categories. If the number of members in this representative group are drawn from the separate categories in proportion to the total number of members in each of those categories, then the group not only represents all types of reaction but also represents them in proportion to their relative weighted importance in the population as a whole. Once this is achieved, the representative group ideally becomes a sample from which the population parameters may be calculated.

Such a sample, like all forms of perfection in imperfect conditions, is virtually an impossibility. It could never be formed in the way described since that would imply a first-hand knowledge of the population and it is this knowledge which samples are expected to supply. Since it is impracticable to identify a sample which exactly represents a population, the difficulty thus posed of selecting adequately representative samples is overcome by making the samples relatively large. As the size of a sample is increased it approaches more closely to the population itself and here, as in other fields of statistical inquiry, there is more safety in large numbers. The law of large numbers argues that the larger the number of observations so the more closely will they tend to be representative of the total population from which they are drawn.

But large samples may be troublesome and costly, and it is necessary to find some way of selecting a sample size which will give estimates of population parameters within specified degrees of possible error. If a margin of ± 3 per cent is permissible in the estimating of such parameters, so that it will make no appreciable material difference to the inquiry on hand, there is nothing to be gained by attempting a greater degree of accuracy. On the other hand, the sample must not be too small.

For any two samples taken from the same population, the probability that the sample statistic for the larger sample will be within a given margin of the population parameter is greater than it is for the smaller sample. The larger the sample, so the less will be the variability in the sample proportions in much the same way as it was seen that, in coin-tossing experiments, the true probability ratio was approached more closely as the number of experiments increased. The variability is reduced because the items causing the variability tend to become relatively less important, as a man who shines in a small group will appear relatively less outstanding in a crowd. If there is a bowl containing one red ball and one black ball, a blindfolded person who selects one ball may find that after 50 selections he may have chosen the red one 30 times. This represents a 60% proportional selection. In the next 5,000 selections, he may select the red ball 2,498 times, or almost 50% of all selections. Then in all of 5,050 selections, he will have selected the red ball 2,528 times, or approximately 50·06%. The effect of the first exceptional sample is swamped by the expected total results of all selections.

The optimum size of a sample in fact depends upon the nature of the population and the nature of the question to which we are seeking an answer. To ascertain what is the optimum sample size for a particular investigation provides one of the most complex of problems with which the statistician has to contend. Contrary to popular belief, the size of a sample does not depend primarily upon the size of the population from which the sample is taken. Instead, the sample size depends upon the degree of accuracy demanded in the representativeness of the sample statistics as an estimate of the population parameter.

It is possible to demonstrate mathematically that, if the population is large, the standard deviation of the distribution of the sample means, which is referred to as the standard error of the mean, is approximately equivalent to the standard deviation of the population divided by the square root of the number of items in each of the samples. That is:

$$\text{standard error of the mean} = \frac{\sigma}{\sqrt{n}}$$

If the distribution of the population conforms to the normal curve, then so will the distribution of the sample means also conform; and, as has already been noted, the latter distribution may be normal even if the population distribution is not. Reference to the properties of the normal distribution then enables an estimate to be made of the accuracy of the sample mean as a reflection of the true population mean, and such estimates are calculated within the confidence limits admitted by the normal distribution. Thus some 95% of all sample means will lie within two standard errors on either side of the true mean (i.e. a total range of four standard errors) so that there is a probability of only about one in twenty that the deviation between a sample mean and the true mean will exceed a value greater than twice the standard error. Similarly it is almost certain (i.e. 99·73%) that the deviation between the sample and population means would not be greater than three times the standard error.

Thus, if in a survey of the weights of 2,500 objects it is found that the mean weight is 155 lb. and that the standard deviation is 25 lb., then the standard error may be calculated

$$\text{s.e.} = \frac{25}{\sqrt{2{,}500}} = 0{\cdot}5 \text{ lb.}$$

Then we may be 95% confident that the true population mean will lie within 2 standard errors on either side of the sample mean; that is, within plus or minus (2 × 0·5) lb. There is thus a probability of 95 in 100 that the true population mean lies between 155 ± 1 lb.; that is, between 154 and 156 lb.

It will be noted that we have substituted the *sample* standard deviation (25 lb.) in place of the *population* standard deviation in the formula. This is simply because we do not know the value of the latter, this being one of the population parameters that we wish to deduce from the sample. We therefore have no alternative but to use the sample standard deviation as an estimator for the population standard deviation. Thus, to obtain one value we have to use another value which is itself a reflection of the first. At first sight this process may appear to be parallel to the antics of the inebriated gentlemen who insisted upon seeing each other home. Its justification, however, is based upon the fact that if a sample is exactly representative of a population then their standard deviations would also be identical. Although no sample will ever attain such perfection of representativeness it may be sufficiently near to perfection to permit an approximate equivalence to the two standard deviations. This assumption may introduce an undesirable element of error, but this will be minimized if the sample is large. To make an estimate of the population standard deviation is not possible merely by inspiration out of thin air and it can only be reasonably based upon the results derived from the sample.

The standard deviation was shown to be 25 lb. derived in a sample size of 2,500 and that the standard error inherent in a sample of this size is only 0·5 lb. Reference has already been made to the confidence limits within which the statistician may express his confidence that the respective means of a sample and the population will be within specified limits of each other. This confidence in the estimated range of possible results can naturally be increased by widening the range. This, however, is achieved at the expense of loss of precision, and it may be that in many surveys the usefulness of possible results will depend upon their falling within a narrower range. It will then be required to reduce the range of possible errors while at the same time retaining the same degree of confidence. The formula for the standard error in the above example was:

$$\text{s.e.} = \frac{\sigma}{\sqrt{n}} = \frac{25 \text{ lb.}}{\sqrt{n}}$$

where n is the sample size. The value of the standard error may therefore be reduced by increasing the size of the sample. For different sample sizes, the respective standard errors, assuming a standard deviation of 25 lb., would be:

n (size of sample)	100	400	900	2,500	10,000
s. error (lb.)	2·5	1·25	0·83	0·5	0·25

It will be noted that to halve the standard error it is necessary to multiply the sample size by four. This is because the standard error varies inversely with the square root of the sample number and not with the absolute sample number.

The confidence limits are based on probability theory and do not exclude the possibility of sample means falling outside the specified deviation limits measured from the parametric mean. They merely indicate the probability that such an event will not occur. Thus, if we are 95% confident that a mean will lie within two stated limits then these are the 95% confidence limits. What is the position if a sample mean is found to fall within the other 5% of non-typical cases? The probability is that this would happen only once in twenty times. How would we know whether the value was non-typical of an assumed population or whether the sample was not in fact drawn from that population? There are certain techniques which may be applied to this problem[1] but one simple course which suggests itself is to take another sample if this is practicable. The probability that both samples will have means outside the confidence limits is

$$\frac{1}{20} \times \frac{1}{20}$$

1. See Appendices 7 and 8.

or once in four hundred times. One single sample may not therefore tell us very much, unless it is sufficiently large, but as statistical surveys become more frequent and as more knowledge is gained from more and more samples of the same population, so the better able are we to apply the fruits of experience to the assessment of the variables involved.

The probabilities so far discussed have been in connexion with sample means. Sampling, however, is very often carried out to ascertain what proportion of a population possesses specified characteristics. For each characteristic, each member of the population must either possess it or not, provided it is a clear-cut concept so that there is no confusion caused by marginal differences. Individuals may then be categorized in definite classifications of the 'haves' and 'have-nots', and calculations may be made to assess the probability that an individual possesses the characteristic. The standard error for a proportion becomes

$$\text{s.e.} = \sqrt{\frac{pq}{n}}$$

where p represents the probability that an individual possesses the characteristic and q represents the probability that he does not. This is another way of saying that p represents the estimated proportion of individuals who possess the characteristic, and q represents the proportion of individuals who do not. Thus if the estimated proportion of individuals possessing the characteristic is 25%, then the standard error (expressed as a percentage) is

$$\sqrt{\frac{25 \times 75}{n}}$$

If the standard error is 2% then

$$2 = \sqrt{\frac{25 \times 75}{n}}$$

whence n is approximately equivalent to 469. A sample size of 469 would therefore be necessary in order that the desired standard error of 2% might be derived from the sample, and we could then be 95% confident that the parametric proportion would lie within the limits of (25 ± 4)%, or between 21 and 29%. Similarly, *n* gives the sample size which, for any estimated population proportion, will give the desired degree of accuracy as measured by the size of the standard error.

Here again it is mainly the desired degree of accuracy required rather than the size of the population which decides the sample size. Nevertheless the relationship between sample size and population size does have some effect, but this is negligible unless the sample represents a large proportion of the population. It is in fact generally ignored in practice unless the sample size represents at least one-tenth of the total population, and is also often ignored unless it represents one-fifth of the population. At this latter level the standard error, unless adjusted, would be overstated by about 11%.[1] A sample size of 1,000, however, will give a standard error value of sufficient accuracy with regard to a population of 50,000, and a similarly sized sample would give an equally accurate result for a population of 100,000.

Subject to due consideration being given to the formula for the generation of optimum sample sizes, the actual size of a sample will very often depend upon the further consideration of hard cash. Surveys are expensive exercises and it is necessary to balance the estimated value of the results desired against the estimated cost of obtaining those results. If only a limited sampling survey is possible within a limited cash structure and this will not give sufficiently accurate results, then either the amount of cash to be made available for the project must be increased or otherwise the project should be abandoned. To conduct a survey which is known from the start to be incapable of giving a desired accuracy merely wastes time and labour which might have been more profitably employed.

The expense of surveys makes it desirable to use the same sample for a number of different though possibly related characteristics. Odhams Press, for instance, carry out a survey of the market for domestic appliances and furniture. They ask each respondent not only whether she has a gas cooker but also whether she has an electric kettle; these are related yet quite

1. See Wallis and Roberts, *Statistics – A New Approach*, Methuen, p. 369.

different characteristics and each has its own pair of positive and negative categories into which respondents may be classified. Within each category, the respondents are further sub-classified according to other characteristics. The owner of an electric kettle, for example, may be asked whether or not she bought it within the last five years and, again, for details of the manufacturer's name.

The respondents to the last two questions in practice form a sample within a sample. Of the original main sample size of 4,160 housewives, only some 1,260 possessed an electric kettle and of these only 469 had purchased one during the previous five years.[1] Consequently the information concerning the fact of ownership of an electric kettle is based upon the full sample of 4,160, whereas information as to kettles purchased during the previous five years is based only on a sub-sample of size 469. The populations are different. The first question as to ownership is designed to give information as to what proportion of households in Great Britain do in fact own an electric kettle; the statistical population is therefore the total number of households in Great Britain. The question as to purchase date, however, is designed to show *inter alia* details of the purchasers; the statistical population in this instance is therefore the population of electric kettle owners. The question as to the brand purchased is designed to show the proportion of the total market enjoyed by the respective manufacturers. Here the emphasis is upon an attribute of the kettle and the relevant population is thus the population of electric kettles, although it is expressed as the population of kettle purchasers.

This sub-classification of a main sample into a number of sub-samples of constituent populations within the main population is a matter of convenience as well as of economy since much of the information derived from the survey may be processed in one major operation. Nevertheless, the procedure has its limits. Each sub-sample must be sufficiently large to ensure adequate accuracy within its own sphere. If there are a number of such sub-samples and no member of one sub-sample is at the same time a member of

1. *Woman and the National Market – 1958*, Odhams Press.

another sub-sample, then the main sample must obviously be large enough to accommodate all the members in all the sub-samples. Thus, if there are ten sub-samples of the following sizes:

A	150	F	1,400
B	1,000	G	1,600
C	1,500	H	500
D	700	J	700
E	800	K	50

then the original sample must be 8,400.

Sub-samples are not necessarily mutually exclusive. An individual who owns a gas cooker may also own an electric kettle. This does not affect the principle, however, that the original sample must be sufficiently large to ensure that the size of the smallest sub-sample is also sufficiently large for its purpose. In the example above it will be seen that a large sample of 8,400 produced a sub-sample (K) with only 50 individuals, and another (A) with only 150 individuals. Whether these samples are large enough is a question of fact taking all the circumstances into consideration. If they are not sufficiently large then the results derived within them will be useless and should certainly not be processed and then presented as if they enjoyed the same degree of accuracy reflected in the information derived from the other sub-samples. In this connexion it is important to maintain a clear image of the inherent standard errors. The degree of error in a sample of 8,400 will be relatively much lower than the degree of error in a sample of 150, and it is therefore not valid to assume that all results derived from a large sample will enjoy the same degree of accuracy, if some of the results are in fact derived from sub-samples within the main sample.

A nominally large sample taken from one population may thus present only a very small sub-sample of certain population groups within the main population. A magazine editor may produce 'evidence' that a particular staff-writer is more popular than another, this inference being based upon a sample of 500 interviews which the editor confidently considered sufficiently large. Of those who expressed an opinion, 50% preferred writer A, whereas only 35% preferred writer B, while 15% could not make up their minds. The editor admits that not everyone expressed an opinion but he is interested in giving the majority of readers the type of article which they prefer; if 50% prefer writer A then surely he is justified in giving A more space.

Whether the editor is right or not depends upon certain data which are not quoted. Suppose, for example, that 250 of the respondents had never read the journal at all and that, of the others, only 100 had read articles by both of the writers. This could be tabulated:

Number who have read A only	75
,, ,, ,, ,, B ,,	75
,, ,, ,, ,, A and B	100
Total number of readers	250
Number of non-readers	250
Total number of interviews	500

Only those who have read both A and B can effectively express a preference and, for this particular purpose, these 100 readers form the sample. The effective sample is not 500, since the other 400 respondents have been ruled out as not belonging to the same population. The total of 500 may be used to provide readership statistics, but it cannot be used as a base for statistics referring only to a sub-group indicating preference. The original statement that 50% of those *expressing an opinion* preferred A may be correct but it is inadmissible to present this statistic as if it were derived from a sample of 500. Presentations of this type, however, do unfortunately occur and may be rendered even more indefensible in such bold statements as '50% of readers prefer writer A' with no reference at all to the fact that the statistic does not refer to all readers but only to those who actually expressed an opinion. There were only 100 of the latter, so that in fact only 50 out of 250 readers really expressed a preference for A.

Confusion between samples could indeed lead to serious errors. The editor, undismayed by his previous efforts, decides to carry out a survey to discover whether detective stories are more popular than western stories. He finds that 51% prefer westerns whereas 49% prefer detective stories. Any majority is good enough for him and he immediately introduces more western stories into his columns. After a few weeks the magazine circulation figures drop away. What has happened? Two things have happened, and the first of these is that the editor has again confused his samples. Suppose the sample data are

Respondents	*Prefer western*	*Prefer detective*	*Total*
Readers of the magazine	50	200	250
Non-readers	205	45	250
Totals	255	245	500

He had noted that 50% of the respondents were readers of his magazine and that 50% were not; this, he was confident, should give a fair picture. But he applied the percentage results from the complete sample to the sub-sample of his own readers. He failed to observe that of his own readers there were 80% who preferred detective stories. Thus, when the magazine reduced its detective story content, 80% of the readers stopped buying the magazine.

The second point of importance is that, quite apart from confusing the sample percentages, the editor has accepted that a definite preference was discernible in the main sample, whereas the figures do not in fact suggest this. These figures are

Prefer westerns	255	51%
„ detectives	245	49%
Total	500	100%

A numerical difference of 2% between the sample proportions (that is, between 49% and 51%) is insufficient to point to a real difference between the related population proportions. Just how slender the sample difference is may be judged from the fact that it represents ten respondents out of five hundred. If only five of these were to have changed their minds the result could have been in equal fifty-fifty proportions.

If in fact 51% is the expected proportion then the standard error of the proportion for a sample size of five hundred is

$$\sqrt{\frac{51 \times 49}{500}} = 2{\cdot}24\% \text{ (approx.)}$$

This standard error implies that, at the 95% confidence level, we can only verify that the population proportion lies within plus or minus 2(2·24)% of the sample proportion – that is between 51 ± 4·48% or roughly between $46\frac{1}{2}\%$ and $55\frac{1}{2}\%$.

A population parameter can only be an estimate if it is derived from a sample statistic. It should always be expressed as lying within a specific range rather than as a specific value. In uncontrolled circumstances such as are involved in the measurement of general population parameters, it is necessary to have comparatively large samples since the standard error reduces as the size of the sample is increased. In some control techniques, however, much smaller samples may be usefully employed. These are discussed in Chapter 20.

17 Populations and Samples

The sample size is largely decided by the degree of accuracy required in relating the sample statistics to the relative population parameters. This is another way of saying that the accuracy depends upon the sample size; this however is merely because, as a random sample is made progressively larger, the sheer weight of numbers reduces the overall variation in the values observed so that the sample more faithfully represents its population. Thus it will be seen, and indeed will probably have been obvious from the beginning, that the accuracy of sample statistics as estimators of population parameters depends upon the faithfulness with which the sample truly represents the population. Weight of numbers, however, provides only one way of achieving this desirable representation. It is not inevitable that samples must be large; other ways have been found of reducing the requisite sample size without sacrificing too much of the accuracy to be expected from the use of larger samples.

The problem of selecting representative samples therefore involves consideration of factors other than mere sample size and it can be a very complex problem indeed. Different methods are called for in dealing with different requirements. If it is desired to ascertain some details of general human behaviour, such as expenditure on food in England, then the sample must represent the population of food-buyers. If, however, it is desired to know something about doctors, such as the number of patients treated each year, then the sample would consist only of doctors representing the population of doctors. In other words, a member of one population cannot be included in a sample to represent another population unless he is also a member of that population.

The evidence of this truth is so fundamental that it should not be necessary to lay emphasis upon it.

It is therefore first necessary to identify the relevant population. A collection of individuals or things may together form a statistical population for one purpose but not for another. The definition of the population of inhabitants of London, for example, depends upon the definition of London's boundaries. The district of Acton in Middlesex is included in the London postal district, and is part of the agglomeration known as Greater London. It is served by London transport services so that its travel problems are integrated with those of London generally. Yet, in spite of these and other affinities, Acton is not within the municipal boundaries of London at all – its administrative parent is Middlesex County Council.

Lack of precision in the practical definition of populations or confusion between different populations may lead to serious error in circumstances which may not always be apparent. One should never accept published sample statistics as representing population parameters without first examining the sample's credentials. It is regrettable, therefore, that despite the stringent requirements of interested professional societies[1] which insist upon the presentation of all relevant sample data, and despite the fulfilment of those requirements by the societies' members, the results of surveys are too often thrown at the public by others without any supporting or qualifying data whatever.

In the previous chapter, mention was made of a confused editor who unwisely applied a sample statistic to the wrong population. The same type of error is often committed in the interpretation of sample statistics. One newspaper report had a bold heading '*Women – One out of every Two gets Backache*'. This conjures up a picture of old ladies hobbling about with their hands significantly placed where the pain hurts most. But look at the picture more closely, leaving aside the fact that it contains no men and the vague suspicion that they may be the cause of all the trouble. Where are all the teenagers and all the other young women? These do not seem to fit in the picture at all and, when you begin

1. e.g., The Market Research Society.

to think about all the ladies of your acquaintance, you soon realize that certainly not every other one is crippled with backache. The statement is clearly a slur against the perfection of the nation's womanhood. What is its evidence and what does it really mean?

Firstly, the category of backache is not satisfactorily defined. Presumably a tired back resulting from overwork would qualify for admission equally with an acute attack of lumbago. The main trouble, however, is with the alleged sample. Looking into the report more closely, one finds the qualifying statement that fifty out of a hundred women attended by a particular doctor admitted to suffering the pangs of backache. All that this sample can tell us is about that particular doctor's patients. He may possibly have a practice in a low-lying and especially damp part of the country where backache, whatever its definition, is more prevalent than elsewhere. In the unlikely event of his patients being typical of all patients throughout the country, then the results of his survey might be applied to all those patients. But this is the limit of their application. The greatest possible population which may be represented by a sample survey of patients is the total population of all patients. The sample cannot be representative of the whole female population of England since this includes women who are not patients. Nevertheless this application is implicit in the bold statement that one out of every two women gets backache.

The report further admits that only some of those reported to have backache actually complained of it; the others confessed to it when specially asked. This surely reveals that the varying degrees of seriousness are important. If a patient who goes to her doctor does not actively complain about backache, it cannot be very serious. The various types of backache afflictions cannot be added together to give any meaningful total, unless at the same time we are given sub-totals for the separate afflictions. There was, indeed, an actual case in 1960 of a woman who reported that she was suffering from backache only to receive the surprise of her life when the doctor told her that she was about to increase her family.

Overriding all this, however, is the fact that women with backache go to their doctors because of that backache and so qualify for questioning, whereas normal healthy women do not go to the doctor and are therefore not questioned. The sample is thus not representative of all women in England; at most it may be representative of all women who are ill; it is possibly only representative of itself in that it represents only the patients of that particular doctor. Perhaps, if that doctor purchased some comfortable chairs and did not detain his patients in the waiting-room for unpleasantly long periods, he might find that the incidence of backache among his patients would decrease to less sensational proportions.

The same report rather naïvely added that about one-third of all patients attending bone and joint clinics also complained of backache. This is scarcely a world-shaking discovery if one considers the proportion of the human skeleton which is represented in its widest definition by what may be construed as the back. The patients with backache complain of it because that is their reason for attending the clinics. To carry the analogy to its absurd conclusion, one might as well report that 100% of all patients in tuberculosis hospitals are suffering from tuberculosis. Of course they are; that is why they are there.

Bias in the structure of a sample must invalidate the use of the sample statistic as an estimator of the population parameter since the population is not accurately represented by the sample. It is a bias which creeps into many samples. Due allowance for it may be made where its existence is realized or where there are other available data which may be compared with the sample statistics. It is much more dangerous when it insinuates itself into surveys the results of which cannot be checked, and it is in order to eliminate bias that the statistician attempts to use random samples. The statistical idea of randomness is more precise than the ordinary dictionary definition of haphazardness, and there are some very odd ideas on the subject. I remember one foreign student on a market research course who, in connexion with a survey into the use of toothpaste, thought that the way to achieve a random sample was to knock on the front door of any house and to thrust a tube of toothpaste into the hands of the surprised occupier. He was thinking in terms of the sample tube of toothpaste instead of a sample of the population of toothpaste users.

A random sample is in fact a sample which is drawn from a population in such a way that every member of the population has an equal chance of selection as a member of the sample, and that his inclusion in or exclusion from the sample could not be affected by any factor other than chance. Randomness therefore applies directly more to the process of selection of the sample than it does to the sample itself. It is haphazard only in that it is indiscriminating – every member has an equal chance of selection. The precautions which are taken to ensure this equality of chance, however, are anything but haphazard. A party of school children may cross a road in haphazard order, but the policeman on duty has to control the party as a whole to protect it from other road users and to avoid accidents. In a similar way the statistician has to control the selection process in order to reduce the possibility of the intrusion of any factor which would destroy the randomness; he tries to eliminate the possible bias.

Randomness is essential in statistical sampling because the pattern of probability naturally depends upon the nature of the population, and only if a sample is random may the mathematical laws of probability be applied to the sampling variabilities. Randomness of selection also implies that the population, from which the sample is to be drawn, is clearly defined since every member of that population is to have an equal chance of selection. How does one ensure that a sample shall be truly random? For comparatively small populations, it is sufficient to assign a serial number to every member and then to draw numbered balls or counters or lottery tickets out of a container. The number on each ticket coincides with the number assigned to a member of the population so that in effect, therefore, the totality of tickets within the container represents the statistical population and the serial numbers of the tickets selected in the draw identify the members of the population who are to comprise the sample.

This method is by no means foolproof. Does every counter or ticket really have the same chance of selection? A folded sweepstake ticket, for example, may depend for its selection primarily upon the particular manner in which it is folded if this differs from the way in which other tickets have been folded. Every counter or ball used must be exactly like its other companions in the container; for otherwise one which is slightly heavier or more slippery might tend to remain at a lower level in the container or might be too difficult to grasp so that it would avoid selection. A parallel may be drawn here to the sampling of paint. In testing paint it is not sufficient merely to test a small amount removed from the surface in a newly opened tin. A certain proportion of the heavier constituents of paint tend to sink to the bottom, so that the whole tinful needs to be thoroughly stirred before being used.

This selection method is indeed not considered statistically satisfactory, although the notion of assigning numbers to members of a population is extensively followed. This notion derived from the necessity of removing all possibility of bias being introduced into a sample by the influences of human preference. A person, who is instructed to test the standard quality of milk in twenty bottles out of every hundred will undoubtedly in time take most of his samples from those bottles which are most easily within his immediate reach so as to save himself the bother of moving all the bottles. Again, a market research interviewer, if left to her own resources, may select out of two possible respondents the person who for some indefinable reason attracts the attention of the interviewer. This preferential selection may be performed quite unwittingly but it is none the less dangerous for that.

Selection therefore ideally needs to be a mechanical rather than a human process. The members are assigned their respective serial numbers and the selection of these numbers must be left to chance. A common procedure nowadays is to use tables of random digits. A number of these have been published[1] and have been subjected to various tests to ensure that there is no apparent bias in the order of the digits. One could, however, build up one's own table by spinning a special ten-sided dice, the sides being numbered from 0 to 9 consecutively, and writing down the results, provided it is possible to eliminate bias from the method of spinning the dice. The published tables of random digits, however, have been shown to be reliable or, at least, have not been shown to be unreliable.

1. For example in *Cambridge Elementary Statistical Tables* (by D. V. Lindley and J. C. P. Miller).

The first step, then, is to assign numbers to all the members of a population. This, in itself, is not always as easy as it sounds. The nominal roll of members of a professional society is often out of date before it is actually published, because of deaths, new elections, and other changes. Secondly, not all of the members may be in practice although they are still members of the society. Any survey of practising professional men would have to take these facts into account. Next, is the population in fact homogeneous? – since only population members who share common relevant characteristics may be compared. Again, is the so-called population in fact complete? That is, does it include all members who, although otherwise unlike, do in fact share a common characteristic which qualifies them for inclusion in the population?

All these points must be sorted out before it is possible even to assign the serial numbers to the population members. There is, indeed, an infinity of populations and, as has already been noted, it is essential that one's spyglass should be trained upon the right population. The next step – that of selecting random digits which shall form the serial numbers to be selected – is very much more simple. They may be read off from a table of digits. The digits are usually printed in blocks of figures so as to make the tables more easily readable than they would be as an undivided mass of digits. These groupings have no statistical importance. The digits might be

49712 61826 01016 90438

The use of combinations of digits depends upon the size of the population. If there are 999 members in the population then every number from 001 to 999 must be included and in order to achieve this the digits must be grouped in threes. The above digits would then provide the numbers

497 126 182 601 016 904

and these might be the first six numbers representing individual members of the sample. Further numbers would be derived from subsequent digits in the same way until sufficient numbers had been selected to complete the sample.

If there are, say, only 825 members in the total population then some of the numbers generated will be greater than 825. It is also possible, whatever the size of the population, that the same number may be generated more than once. These inconveniences are ignored and we proceed instead to the next generated number. This could be a tiresome business if the population were only, say, 120, since about 88% of all numbers generated would probably exceed that number. The process may, however, be rendered less tedious by arranging for more than one generated number to correspond with the assigned number of each population member. Thus the generated numbers 120, 270, 420, 570, 720, and 870 may each be made to correspond with the assigned member number of 120; then the numbers 001, 151, 301, 451, 601, and 751 will each correspond with the first numbered member of the population list. This will still leave some generated numbers which will have to be discarded, but the number of such rejections will be reduced considerably.

Where populations are really large the whole method of assigning population numbers is of no practical use. To obtain a random sample of the households in the United Kingdom by this method it would be necessary to number every household listed in the electoral register. Some other more practical methods are required and, of these, systematic or quasi-random sampling is often employed. In this, every *n*th member of the population is selected. A starting place is chosen at random on the population list and, if there are 100,000 members of the population and a sample of 1,000 is required, then every hundredth member may be selected. It is sometimes complained that this is not strictly a true random process since every member after the first is pre-selected according to the stipulated interval of selection. Once the first member has been chosen, the other members of the population do not all have the same chance of selection. They either join or do not join the sample in strict conformity with the requirement that every hundredth member shall be in the sample and that all the others will be excluded. Selection is thus not a matter of chance at all.

Against this view it is also argued that, although 99 members of the sample are selected by reference to their listed relation to the other member and although their identities do not become known until after the first member has been chosen, nevertheless the 100 members are in effect all selected in one operation. There is an apparent paradox here in that, although they are selected as individuals, they are nevertheless identified as a group of individuals by virtue of their relative list positions. This group is in fact chosen as an entity; it is merely identified by the first member and is not a consequence of his having been selected. Any spaced group could have been chosen so that if the first member is identified by a chance decision then all groups and therefore all individuals have an equal chance of selection. This statement, however, cloaks a significant fact. A truly random sample size of one thousand could theoretically be formed by the combination of any thousand members of the population and any one or more members could appear in more than one sample. Mr Smith might appear in sample number one in company with Mr Brown, and he might also appear in sample number two in company with Mr Black. In quasi-random samplings, however, there are many fewer possible different samples. Mr Brown would appear in a sample with Mr Smith only if there was the stipulated interval between their names, and Mr Smith would not then be free to appear in a different sample with Mr Black. Mr Smith's fellow sample-members would always be the same individuals in the short term, subject to changes being made in the list and minor marginal differences. This difficulty can of course be overcome by changing the size of the interval between selected names for different samples.

Quasi-random sampling of this nature would thus be unsuitable for small populations or in circumstances where the sample represents a relatively high proportion of the population, since there would be the risk of virtually the same sample being selected too often. Where the population is large, however, the method approximates sufficiently closely to a true random process, provided that the list used is not arranged in such a way as to give a particular type of population member a greater chance of occurring at the same regular interval as that laid down for the sample. In an extreme case, for instance, where there are nine men and one corporal in each hut on an army camp and their names are listed in hut order but with the corporal's name first, then one of the samples taken at intervals of ten names would include all the corporals. Where all the members of a sample consist of a particular type of member in the population, their combined presence, to the exclusion of other types, must introduce bias into the sample results if the difference in character between the types of member is relevant to the purpose of the survey for which the sample is drawn.

The list chosen as the sampling frame must contain the whole population. There is no better illustration of this requirement than that provided by the terrible mistake made in 1936 by an American magazine which forecast that the Republicans would sweep all before them in the Presidential elections. A huge sample of 10 million people was selected and, of these, over 2 million replied to the questionnaire. The size of the sample, coupled with the fact that the same sample had previously been used successfully in four earlier elections, was taken as a sure guide to the outcome of the 1936 election. The sample returns indicated that Franklin D. Roosevelt would be defeated whereas, in fact, he was elected with one of the largest majorities ever recorded in American history. The apparent reason for the failure to make a correct prediction was the fact that the sample chosen was not representative of the total population of American voters. The magazine's inquiries had been addressed to its own readers and also to telephone subscribers as listed in the telephone directories, so that the sample represented only the population of people who either owned telephones or were readers of the magazine. Consequently it did not represent the mass of voters who neither read the magazine nor owned a telephone.

In the particular circumstances of the country's economy at the time of the election, it appeared that, by choosing a sample from people who could afford to have a telephone installation, the organizers of the survey were in effect selecting people of preponderantly one party. In some indirect way, therefore, the possession of a telephone was at that time enjoyed mostly by people whose political views largely favoured one party. More Republicans than Democrats possessed telephones and the results were therefore weighted in favour of the former. This state of affairs had apparently not existed at the time of the earlier elections. The fact that the sampling frame had been used a number of times previously with apparent success did not save the survey any more than did the huge size of the sample. Size alone is not enough. It is a common misconception that if a sample is large then the necessity for randomness in sampling is proportionally lower. The 1936 election results proved that this is not true.

The error in interpreting the survey results may also have been due in some measure to the comparatively low voting rates in America. The percentages of eligible voters who actually record their votes in American Presidential elections are consistently much lower than the percentages recorded in British and other European elections. In the 1948 election, for example, only 50·1% of the electorate voted, while in the 1952 and 1956 elections, the percentages were 63% and 60·4% respectively.[1] These are very low percentages and, whatever sampling frame was employed, a high proportion of people interviewed might record their opinions as to what the result might be even if they themselves did not intend to vote. While this might be reliable as reflecting the relative popularity of a party, this could not be translated into terms of probable voting results.

A modified form of random sampling, referred to as stratified sampling, is rendered necessary when the population is not homogeneous. Such a population might be the total demographic population of the United Kingdom, since this is made up of men and women in different age groups, in different economic and social circumstances, in different trades or professions, and with widely differing views on practically every subject. They all belong to the one total population and share some common relative characteristic (e.g. that they all drink milk) but are divided into groups according to some related degree of affinity to the characteristic (e.g. members of one group may drink more coffee than tea and thus consume greater quantities of milk per person). When the various groups of strata have been identified, a simple random sample is taken from each, and all the samples are then incorporated in the total result by some weighting method in proportion to the relative size of each group.

The necessity for stratification would seem to run counter to

1. *The Economist*, 24 September 1960.

the contention that a simple random sample is really representative of the whole population since, if it is thus representative, then the stratification should occur in the random sample itself. Positive stratification, however, is much more satisfactory. All sample selection is subject to some error but if the population can first be classified into strata, then this method ensures that the members of the different strata are included in the sample in proportion to their relative numerical importance in the total population. Stratified sampling will of course be subject to error if the weighting of the various strata is inaccurate. For some large populations, the basis of such weighting may be largely guesswork and it is important that all results derived from stratified samples should be assessed with this possibility well in mind.

Another form of sampling is that of multi-stage or cluster sampling. For this the population is divided into counties or other large sampling units and a random sample of smaller units is drawn. The total number of necessary interviews is then allocated between the main unit areas according to their relative population sizes. In each main area the number of interviews is then subdivided between urban and rural districts upon the same principle of proportion. The interviews for urban districts are allocated between various towns, although there must be a restriction here since not every town can have interviews allotted to it even if there are enough interviews to go round. The cost of conducting interviews in every single town would be prohibitive. Finally, in the selected towns, the actual respondents to be interviewed are chosen from the electoral registers or by some other convenient method.

By this method the samples from each population unit are proportionate to the size of the latter. Some modern multi-stage methods, however, do not select units at the appropriate stages by strictly random methods. Instead, they select the units with selection probabilities proportionate to their individual sizes since they will rarely if ever be of exactly the same size. A county which has twice as great a population as another will then have twice as many chances of being selected. Then, instead of having different size samples for the various counties, each sample will be of exactly the same size so that the lower probability of the selection of an individual in a large population unit just sufficiently offsets the greater probability of that unit having been chosen. This restores the overall equality of chance of selection to each and every individual.

Each of these methods depends upon the assumption that two members of a population selected in one town will give as good a result as will one member in each of two towns. This in general tends to be true as the number of interviews is increased; individual differences tend to cancel each other out when the stage results are combined to produce the results for the complete sample.

Quota sampling also depends upon a similar kind of assumption. This form of sampling is adopted to overcome some of the problems posed in the correct functioning of random sampling schemes. Interviewers are not given lists of people whom they are to interview but, instead, they are given details of how many people they should interview in terms of age range, sex, social, and other classifications. In effect, therefore, the total population is first stratified but, instead of a random sample being taken within the strata, it is left to the interviewers to select their own respondents.

This at once admits the possibility of errors as a result of the intrusion of the human element into our calculations, even though interviewers may be given strict instructions as to how they are to make their selections. Instructions can never be sufficiently strict to cater for all the possible deviations, I was once interviewed in a London park during the summer holidays. The weather was gloriously sunny and the interviewer probably felt that everyone and his wife would be out in the sunshine. The park was the obvious place in which to look for them and the hot pavements of nearby streets were too uncomfortable anyway. This interviewer may have obtained a good sample or she may not have done so. If her instructions merely required that she should interview a specified number of men and women in special age groups, she could easily have found the right numbers in the park. But, to state the obvious, her sample would consist only of people who were in the park – a special class of individuals who were on holiday or who lived nearby. People who were at work or who lived far away from the park or who, because of age or family commitments, could rarely find time to go to the park would not be represented in the sample.

In practice, however, the selection of suitable candidates for interviews is normally much more precisely defined and controlled by the central agency in charge of the survey. Interviewers may be instructed as to exactly where and when they must carry out the interviews and also as to the type of person they are to interview. In addition to selecting respondents who shall fit into certain age and sex categories, the interviewer may also be instructed that a specified proportion must be office workers or shop workers or that the respondents should also appear to conform to certain other category characteristics. Human failings tend to recur, and a wide-awake research manager should soon be able to recognize any recurrent signs of bias in a particular interviewer's work. Checks may be carried out by referring back to the respondent, and questionnaires may be worded in such a way that they are self-checking not only upon the respondent's replies but also upon the interviewer's recording of those replies.

Nevertheless quota sampling, despite all these checks, must retain a number of possible defects. The human bias cannot be entirely removed and the sample is not strictly a random one. Yet it is a very common form of sampling and is considered to provide sufficiently representative results in surveys where the highest degree of accuracy is not essential. If, indeed, the human bias element may be virtually eliminated, so that quota sampling more closely approximates to the equivalence of ordinary stratified random sampling, then its advantage is that it can achieve much the same result more easily and cheaply.

In random sampling, interviewers are detailed to obtain their interviews at specified addresses or with specified individuals. Because of the rigidity of these requirements, an interviewer must sometimes make a number of calls before finding anyone at home at some of the houses on her list or before locating particular individuals. If she could organize her tour of duty so that she would not need to cross her own path, she could complete her calls in much less time than it takes her in practice when she must forever be retracing her steps to call back at certain addresses.

Yet, in theory, it is necessary to ensure that individuals selected by random methods should in fact actually be interviewed, for otherwise the basic idea of the method of selection is not realized. It is no solution to knock next door and obtain an interview there. A household where both husband and wife are out at work during the daytime, and therefore not available for interview, might probably provide different opinions or information than would their next-door neighbour who does not need to go out to work. In order to overcome this difficulty, some surveys compromise by providing that, where an interview cannot be effected, the interviewer should then proceed to substitute some other alternative individual who has previously been selected by random methods.

Some doubt exists, however, as to the relative importance of the individual omitted, because he is not available, compared with the individual substituted. If people refuse to be interviewed there may be a good reason for this which is relevant to the inquiry. If, for example, a Labour Party canvasser were to ask a Conservative supporter to supply precise details of his political leanings, the latter would probably refuse to do so. The interviewer might then have to proceed down his list of substitutes until he reaches a Labour supporter who is prepared to give an answer. Bias in the replies received would be obvious. It is true that interviewers normally preface their requests for information with the assurance that they, the interviewers, are quite impartial to the subject matter discussed, but there may be some less obvious reasons why an individual might not wish to disclose his views. Yet it might well be that his views, if known, could in conjunction with those of other similar individuals make a noticeable impact on the results.

However, provided the substitutes are selected at the same time and by the same method as are the rest of the sample, then there seems to be no valid reason for their exclusion even if there is really no valid justification for their inclusion. If a sample size of 500 is required, then a total of some 600 might in fact be selected, those in the last hundred being included in the actual sample only to the extent that refusals are experienced in respect of the individuals in the first five hundred. This may not give perfectly random results but, unless the population size is an

18 Popping the Question

The most appropriate form of sample having been decided upon and the actual sample having been selected, all that is then necessary is to bombard its members with questions in the hope of receiving intelligent replies. That is all! It sounds so simple but it is a deceptive simplicity and this particular aspect of a survey is fraught with as many, if not more, difficulties as are encountered anywhere else in the sampling process.

The procedure is theoretically not unlike criminal court procedure when, the police having brought forward a number of witnesses, counsel set about finding out the truth by asking pertinent and sometimes personal questions. In a survey, the members of the sample are the witnesses; their answers provide the evidence and the tabulation of their answers constitutes the summing-up. The legal system, however, has many safeguards designed to ensure that truth shall prevail. After the presentation of the 'facts' by counsel and after the judge's summing-up, it is still left to a jury to decide the result of the case. With published statistics the reader has to be his own juror. In court cases the jury is present right through the whole trial and it is therefore easier for them to relate each of the stages of the trial to other stages and so to obtain a more comprehensive picture of the whole set of circumstances involved in the case. The reader of statistics is not in the same fortunate position. As has already been noted, published statistics are often thrust at the reader without any supporting data. Before he accepts the value of the answers thus presented, the reader should at least ascertain what the questions were.

The general nature of the questions to be asked will of course be decided in relation to the subject of the survey and the type of answers required. It may seem too obvious to iterate that this implies that, before attempting a survey, one must be absolutely clear as to what information is required. But the obvious cannot be stated too often. If the researcher is not really clear as to exactly what information he requires, how can he expect others to know? Furthermore, only the desired or closely related information should be collected, for otherwise the whole survey may be bogged down under a mass of useless information so that the researcher will not be able to see the facts for the files.

Special attention to the phrasing of the questions is absolutely essential. Silly questions, it is said, receive silly replies. And they do! Additionally, when one deals habitually with the public, it is soon borne in upon one's consciousness that it is not even necessary to ask silly questions in order to get silly answers. The latter come of their own accord. The answer is very often a lemon. In one survey into the use of shaving preparations, a respondent was asked, 'What do you use for shaving?', the expected answer being a reference to some kind of toilet preparation. In a broad Irish accent he announced that he 'always used boilin' hot water!' The question itself was at fault since it did not make clear to the person interviewed just what it was asking about. It could just as easily have been referring to the type of razor used.

The main essential is that the respondent should clearly understand the question. Questions should therefore be short and very much to the point, employing the simplest words possible. But to understand is not enough. If the survey is to be successful, each and every respondent should interpret the question in exactly the same way, and this interpretation must be exactly the same as that intended by the researcher. The wording therefore needs to be unambiguous. When a young suitor pops the question and asks his girl-friend if she will become his wife, she should not have to sit back and reflect 'What does he mean by that?' although she might later wish that she had done so. Yet, try as we may, it is not always possible to achieve the eminently desirable lack of ambiguity. What is unambiguous to a reasonable intelligent person may still give trouble to another individual. Even the simplest phrases can be either misunderstood or interpreted in different ways, and the use of vague adjectives such as *large*, *small*, *cheap*, and *dear* should be avoided unless qualified by some degree of measurement of size or price respectively.

A survey was carried out in New York City in 1943 to ascertain in how many different ways a question might be interpreted.[1] The question had previously been the subject of a national coverage Gallup Poll and asked: 'After the war, would you like to see many changes or reforms made in the United States, or would you rather have the country remain pretty much the way it was before the war?' One hundred and fourteen interviews were carried out and on the basis of the replies received as to specific reforms which were mentioned in the replies, the authors of a report[1] on the survey deemed it possible to rate the respondents as to their respective frames of reference. From these ratings it appeared that about one-third of the respondents interpreted in frames of reference other than that of 'Domestic reforms' which was, presumably, the frame of reference intended in the question.

The importance of correct interpretation is exemplified by the fact that, if a respondent misunderstands a question, he will give an answer to the question as *he* understands it and, in these circumstances, his answer cannot be other than misleading. There is more danger of this happening when the respondent is left to himself in the completing of a questionnaire. If an interviewer is present, she may perhaps be able to make reasonably sure that the respondent does understand the question correctly. Even so, this is not infallible with the simplest of questions. The American Census Bureau, for instance, find that, when householders give details of persons resident with them, some of them overlook the old people and babies. These categories tend to get lost in the oddest manner. Sometimes they are actually overlooked, but in other cases they are deliberately excluded for a variety of reasons. Grandfather is left off the form because he is retired; the new baby who arrived yesterday is not identified as a *person*, this latter description appearing to carry an adult connotation, and no mention is made of him either.

In addition to the necessity that a simple question shall be

1. R. S. Crutchfield and D. A. Gordon, 'Variations in Respondents' Interpretations', *International Journal of Opinion and Attitude Research, 1947.*

capable of receiving a simple answer, it must also be of such a nature that the respondent really is able to answer it. That is, in addition to undertanding the question easily, he must also have the necessary knowledge which will enable him to give an answer. The latter will be useless if it is not based on real knowledge. Indefinite answers resulting from guesses about other people's activities should be counted as 'don't-know'. It is probably not much use, for example, to ask a miner what he thinks about nuclear engineering, but it is not only the highly technical subjects about which a man may be ignorant. He may not even know what brand of soap powder his wife uses. Similarly, many housewives may think they know more about their husbands' expenditure than in fact they do! The first question should therefore always establish whether or not the respondent has any knowledge of the subject.

Furthermore the questions must be of such a nature that the respondents are prepared to answer them. This applies particularly to the more intimate personal details. Extroverts might probably be quite pleased to answer questions whereas more reserved people would not be so forthcoming with their information. A strong bias could develop in the results, and the latter could be properly applied only to the population of individuals with characteristics approximating to those of the willing respondents. They would not be applicable to the population of a country as a whole. Some personal details are required in order to check whether the interviews in total do actually represent the relevant population. Questions of a confidential nature, which a respondent might be reluctant to answer (as to age, for example), can be sprung upon him or her when it is least expected after the respondent's confidence has been won on some easy general questions of a less personal tone.

The positioning of questions within the framework of a questionnaire is a most important consideration. Where possible the questions should lead the respondent along gently, step by step. A logical order in the sequence of questions will help a respondent to marshal his own thoughts and so probably to render his replies more intelligent and consistent. The order of questions can have a number of subtle influences. If a person gives a slightly inaccurate reply to one question, he is not likely at a later stage to admit his earlier error; neither will he give a true answer to a later question if he considers that the two answers would be obviously inconsistent. Nevertheless, it is useful to include self-checking questions at different stages of the questionnaire.

Inconsistencies in a response can easily be detected, but it is not always possible to decide which of the inconsistent pair of responses is correct and which is incorrect; this really invalidates all the replies involved. Much more serious, however, are inaccuracies which cannot be detected since they are not covered by any form of self-checking. It is, regrettably, a fact that respondents do not always tell the truth, either whole or otherwise. This is not necessarily a matter of inveterate lying and may consist of white lies and those classified euphemistically as mental reservations. But if they result in inaccuracies creeping into our calculations, the depth or motive of falsehood is relatively unimportant. A response may either be accepted or rejected. There is no satisfactory half-truth. You cannot add half-truths to truths and get anything other than total confusion.

Respondents may be particularly reticent about their age, occupation, salary, and marital status. These, too, are subjects on which some people are inclined to exaggerate, either for the pleasure of boasting or to conceal their shame. A man who considers himself a failure and is living on a low salary may be too ashamed to reveal his true income. There may be other subtle reasons. It has been suggested that a woman who is separated from her husband will still rightly declare herself as married, particularly if the interviewer is also wearing a wedding ring, whereas the husband is more likely to declare himself single again! How else, it is asked, may we explain that there are always more married women than there are married men? Published statistics[1] show the following details taken from census returns in England and Wales.

	1931 *Census*	1951 *Census*
Married females (thousands)	8,604	11,092
Married males („)	8,490	10,995

1. *Annual Abstract of Statistics* (Table 13) 1959, H.M.S.O.

Similar differences occur in respect of Scotland and Northern Ireland, and it should be mentioned that divorced and widowed persons are excluded from these figures. These figures, however, do not prove the point, for there are many possible reasons for the difference arising out of the specific conditions and definitions adopted for the purpose of the census. They include only those persons actually in the country at the time of the census and thus exclude all servicemen serving abroad.

Whatever the truth may be about respondents' reactions to questions about their marital status, it is doubtful what, if any, reliance may be placed upon personal income statistics derived from taxation sources, although one may at least be sure as to the direction of the bias. How truly can such statistics reflect individual incomes when it is a popular and conventional sport on the part of the individual not to report accurate figures? Despite the many printed warnings as to the outcome of such entertaining practices, taxpayers often attack their tax forms in the sporting spirit of whether they can beat the tax inspector at his own game. This is not confined to business tycoons who find it profitable to employ experts to find legal loopholes so that taxation may be avoided; it involves also thousands of ordinary workers who do odd jobs for cash. An Inland Revenue office is a battleground between authority and the individual; many people regard income tax as an imposition and yearn only to emulate those who can avoid it.

Ordinary surveys will not normally reveal such positive mendacity, but they nevertheless suffer the weakness that they are liable to much the same form of abuse. Taxation authorities have special investigators to check back upon an individual's tax liability. Researchers have no such supporting organization; self-checking questions are very useful but even they cannot reveal wilful untruths if the latter are consistent within themselves. As in matters of taxation, a respondent cannot be relied upon to give a whole truth in any matter where he feels that his personal interests may be affected by his disclosure. This is perhaps one very good reason why husbands and wives should not be questioned in each other's presence!

People when interviewed tend to claim that they act conventionally within accepted or supposed codes of conduct.

Similarly, they may well claim to prefer better quality goods even though they habitually buy goods of a poorer quality, and they may sometimes claim that they do actually purchase the better quality goods. Responses really require interpretation by a process of psycho-logic, a peculiar admixture of logic, psychology, and perhaps psychiatry and psycho-analysis. Interpretation of the responses may, however, be simplified by the re-phrasing of questions; the same question expressed in different ways with different degrees of emphasis may produce quite different responses. An indirect question may sometimes provide a more reliable answer. Respondents may be asked, not whether they have taken some specified action, but instead when they did it. It may be noted that children, not being bound by convention, are inclined to be more honest in their replies to questions. After a film about how to prevent home burglaries had been shown to a group of school children they were asked which character in the film they had liked most. More than 90% voted for the burglar![1]

If you ask someone whether he takes a bath daily, and if he thinks that he ought to do so, or that you think he ought to do so, then he will say that he does have a daily bath even if he does not. The researcher is continually confronted by the fact that responses usually record only what the respondent, once he has been put on his mettle, says that he does rather than what he actually does do. The respondent is encouraged to think that he ought to take a daily bath, since he will reason that otherwise he would not have been asked the question. He is not then likely to let the interviewer discover any shortcomings in his toilet habits. The question may thus provide its own answer, irrespective of the actual circumstances. If it is desired to ascertain facts about bathing habits, it might be better to ask the respondent how often he takes a bath. The question does not then beg a particular response, and there may be fewer claims to daily bathing. Even this device will not ensure complete accuracy, since cleanliness is an advertised virtue and respondents would prefer to be thought of as possessing clean and regular habits than otherwise.

In general, therefore, it is desirable that, although he should

1. United Commercial Bank *Fortnightly Letter*, July 1960.

clearly understand each question, the respondent should not be able to divine the precise purpose of the survey. A retail trader, who pays a trade association subscription according to a scale based upon the number of operatives employed, may return a lower figure, in order to qualify for a lower subscription, by quibbling over terminological differences in the meaning of the word 'operatives'. Is he to include only those actually employed in the particular trade (where the trader carries on more than one trade) and, if so, how is he to classify the office clerks who cope with all the different trades? On the other hand, the trader might for prestige purposes wish to show a high figure for staff and he will then include everybody including his wife and also the occasional window-cleaner. Yet all these figures will some day be extracted from the forms, and tabulated to show numbers of employees per trader and other comparative presentations.

A great deal depends upon the respondent knowing neither the real reason for the survey nor the identity of the sponsors. Interviewers need to dissociate themselves from the inquiry itself and to make it clear that they are merely acting as clerks to assist the respondent in completing the questionnaire. This is not completely possible. The interviewer and the respondent are human beings and it is impossible to prevent their personalities reacting upon each other. An interviewer knows that she must endeavour to treat all respondents alike; it is her job to do so. The willing respondent regards the interview in quite a different light. It is still an event in most people's lives to be interviewed. There is a certain pleasure to be derived from the apparent distinction of being selected from the crowd. Anyone who has conducted a number of interviews will recognize the flattered reaction of many respondents. The majority of people like to be asked to join in an experiment and, in return, also like to be helpful. But helpful people can be a menace to statistics. They sometimes try too hard. In a survey into smoking habits,[1] members of the sample were asked to collect the stubs from their cigarettes and to send them to the research centre. Some non-smokers were so keen to help that they collected cigarette stubs from other people!

1. P. G. and E. A. Parr, 'The Length of Cigarette Stubs', *Applied Statistics* VIII, No. 2. June 1959.

It is not uncommon for male respondents to be particularly susceptible to the charms and general appearance of an interviewer, and their reactions will have very definite influences upon the replies. Disinterest in the subject of a survey may be countered by the pleasure involved in participating in an interview, but irresponsibility cannot be satisfactorily controlled. Not only may a person tend to claim that his actions conform to what he thinks he ought to do; he may also tend to claim that he conforms to what the interviewer might expect him to do as well. A bashful person, too, may give the first answer to come into his head, without thinking seriously about the question, merely in order to conclude the interview as soon as possible. Bashfulness may be rare today, but many other people do give answers without having given them much thought. There are, indeed, many difficulties in the way of obtaining absolutely honest and correct results. Even hypnotism has been considered as one means of overcoming them and baring truths, but this has most properly been proscribed by the Market Research Society.

Many questions cannot indeed be answered by a direct affirmative or negative. A respondent may not be able to think of a suitable response to such questions quickly and this may produce too many 'don't know' answers which will reduce the value of the interviews. In such circumstances it is often usual for interviewers to prompt by asking supplementary questions or by indicating possible answers. This involves the risk that the prompting may be attempted prematurely so that the respondent is led to a reply which he might not otherwise have given. Lists of possible responses may be shown to the respondent, such as a list of magazines in respect of which he is asked to indicate which magazines he has read. It is possible he may indicate at least one of the magazines even if in fact he has never read any of them; and if the list is too long, it may be found that the magazines printed most prominently are more likely to be selected. In order to check the accuracy on this point, it is desirable to ask supplementary questions contingent upon the selection made as, for example, asking which features are preferred in the magazines selected.

Readership surveys of magazines are common and it is not unknown for one magazine to be voted more popular than another, even though its known circulation is far below that of the other. Respondents will also sometimes mention a particular make of product which first comes to mind, particularly where the maker has a widely publicized brand name. This is especially evident in circumstances where the brand name has become popularly confused with the identity of the product to which it is applied. For years the brand name 'Hoover' was synonymous in the public mind with 'vacuum cleaner'. Many people, if then asked whether they had a Hoover would have interpreted the question as meaning 'do you have a vacuum cleaner?' and would have answered that they did possess one although their vacuum cleaner might have been of a different manufacture altogether.

Unreliable responses may also be expected if they depend upon the respondent's memory. The questions are important to the researcher, but they are not of equal importance to the respondent. An individual will normally memorize with an approach to clarity only those occurrences which either interest him or which make an impact in some other way. Memory is not always to be trusted on subjects or events which are of little direct interest to the respondent or which call for a positive effort of scouring the memory track. Many people often cannot recall to mind clear images of events which happened yesterday, let alone last week or last year. In one survey, members of the public were invited to state the length of time which had elapsed since they had last bought a particular household appliance, – was it one, two, three, or four years ago? It is extremely doubtful whether anyone would be able to remember exactly when they purchased an appliance if the purchase was made more than two years ago. The responses obtained would be valuable only if they could be checked against some documentary evidence.

The interpretation of a response is equally as important as the respondent's interpretation of the question. The latter may be specifically designed to facilitate its interpretation, but the response is not subject to any similar form of control. If the response is not a direct affirmative or negative, then its classification, or coding, depends upon an understanding of the respondent's words. If the coding is effected at the survey central control, then the written record of the response must stand alone.

The tone of the response cannot be gauged, but the manner of making the response could in some circumstances be almost as important as the response itself. Another difficulty is that, if the respondent is permitted to speak at length, his response will almost certainly not be recorded in full. If the interviewer summarizes it, then in the final coding it is the interviewer's interpretation and not the original response which is being coded.

The actual coding needs to be as exhaustive as possible in order to ensure that every response will fit into one of the categories. Subject to the provision that there should not be too many subdivisions, the main response categories should themselves be subdivided to allow for the segregation of responses revealing different degrees of reaction. It is also desirable, but not always practicable, that categories should be mutually exclusive, so that each response may be classified in only one category. Some questions, however, seek to ascertain an individual's likes and dislikes. If, in a sport survey, a respondent says that he likes cricket and football, he must be included in each category if the survey is to show how many people like football and how many people like cricket. The classification of responses in this way creates some difficulty in that it disturbs direct proportional comparisons between the categories; the total number of categorizations will exceed the number of answers and the latter cannot then very well be expressed as percentages of the total. Multi-classification also gives undue weight to the responses of the more voluble respondent who states a number of reasons for his reactions if each of these reasons is classified. Correct coding is a very difficult procedure. It should not be left to individual interviewers, for different interviewers might treat the same response differently.

It is apparent that interviewing is a vital stage in a survey and that the utmost possible control must be exercised by the central authority. Complete control is not practicable, but the instructions issued to interviewers must be sufficiently stringent to reduce the error potential. Instructions as to procedure, however, are not enough. Whereas the respondent should not have knowledge of the basic reasons for the survey, the interviewers should have a clear knowledge of the purpose of the investigation and should be thoroughly conversant with the nature of the difficulties which might be expected. They must also have a full appreciation of the necessity for neutrality and impartiality in their dealings with the respondents, and for the ability of placing the latter entirely at ease.

Interviewing is not an easy task if carried out conscientiously. It can be very exasperating indeed. An interviewer's patience may suffer if she meets too many people who cannot make up their minds, and promptings may be given too easily as a result. Asking the same question again and again becomes boring over any lengthy period and a certain amount of nervous strain and fatigue may be experienced. There is a wayward tendency, after repeating the question a number of times, for the interviewer to vary the wording by paraphrasing the question. Interviewers may become too conversant with the question, and familiarity breeds carelessness if not contempt. The length of the interview is of importance also. The shorter it is, so the more quickly may the interviewer complete the task; and it must also be remembered that few respondents, perhaps approached during their lunch hour, can afford to sacrifice too much time in the interests of statistical research.

Human errors at the interviewing stage are by far the most serious of all errors arising during a survey, for they cannot always be detected. Pilot surveys are therefore desirable, before any large-scale survey is attempted, in order to test the reactions of a relatively small number of people. With all the possibilities clearly in mind of the inevitability of mistakes, disinterest, and deliberate falsehoods creeping in at the interviewing stage, some method is obviously needed to assist in the assessment of the final survey results. The statistics themselves cannot give any indication of their own accuracy and there is no way in which the problem may be tackled satisfactorily by mathematics. Instead, one must look back to the beginning of the survey. Notes, which accompany the published figures, should explain how the interviewers were briefed and should also contain a copy of the whole questionnaire. This will enable the reader to study the actual questions and to judge which, if any, might have been capable of misinterpretation or which might have begged certain responses.

It is a useful discipline, before accepting any statistics derived from a survey, to answer all the questions oneself. This will throw both the questions and the answers into their correct perspective. It is also advisable to check whether all the relevant questions were included in the questionnaire or whether any, which might have had a material effect on the results of the survey, were overlooked. The notes should also include details of the categories of individuals questioned and the number of individuals in each category. Statistics which are published without this information should be treated with the greatest reserve. They are not necessarily useless because they are not accompanied by the relevant information, but they should not be accepted as useful until such information is forthcoming.

Yet, in spite of all the difficulties involved, surveys do give very useful results where a range of values, rather than specific values, are sufficient for the intended purpose. A survey, which seems to show that between 50% and 60% of a population have a particular characteristic, is more reasonable than one which claims that the correct proportion is 55%. At this level, random sampling errors alone may account for a range of 2% error (that is ±1%) even for a sample size of 2,500. And, for all its weaknesses, the direct interviewing method is on balance more satisfactory than the postal method in which questionnaires are posted to selected individuals.

The postal method begins by being more scientific since, if properly carried out, the selection is truly random, whereas when the final selection is left to an interviewer it is not perfectly random. But it is no use having a random sample if the persons selected will not return the questionnaire. Few people enjoy filling in forms unless they are really personally interested in the subject of the survey or unless they are those others who must be facetious at all costs and who return highly amusing but otherwise useless concoctions.

Less than five per cent of all persons approached will refuse to be interviewed, but the researcher will be lucky if he gets more than twenty per cent of the members of a postal survey to reply. The response rate will, indeed, often be less than this, although a better response rate may be experienced if the sample is relatively small so that it is a practicable proposition to progress-chase the sample members. A useful rate of some 80% will defeat the whole object of trying to obtain a representative sample, and bias may well become so great an influence that it completely upsets the niceties of balance originally built in to the sample. An even greater refusal rate may be encountered when extra work is required of the respondents in the keeping of records over a period of time. Unless the return of a majority of postal questionnaires can be guaranteed it is not a really worthwhile exercise, and it is an unfortunate fact that this guarantee is only possible if the completion of the forms is compulsory – and only the government has powers of this nature.

Interviewing has other advantages. The more awkward questions may be left until the end of the interview. Even if a respondent refuses to answer these, the interviewer will already have received answers to the earlier and possibly more important questions. In a postal survey, however, the whole questionnaire will be torn up and thrown away if the respondent considers just one of the questions to be too personal. The answers to all the questions are then lost. It is of course important in all kinds of survey that the supporting notes should give full details of the response refusal rate both as to the questionnaire as a whole and also as to individual questions.

Newspapers do not assist in the interpretation of results, for they invariably select what appear to be the most significant points and emphasize these at the expense of tabulations and supporting notes. The latter are rarely, if ever, published. This in itself is not a criticism of the Press since it is their function to draw attention to occurrences and not necessarily to describe them in full. No newspaper reader would wish to see long statistical tabulations to the exclusion of his favourite features, but he should resist the tendency to accept the superficial report at face value. It is because the reports are so superficial that the reader needs to look elsewhere for the real facts of the matter. Some reports do, however, actually misinterpret and emphasize quite erroneous conclusions, thus helping to bring statistics into unjustified disrepute.

With the developing popularization of statistics and psychology have come other forms of journalistic questionnaires to replace the old and disfavoured processes of numerology. This form of what may perhaps be termed statisticology, seeks to demonstrate to the reader such hidden secrets as whether he is a good mixer or whether he really loves his wife or his job or possibly someone else's wife. Points are awarded for each answer to questions and the total number of points scored by a reader are alleged to prove something or other which is often quite different from the stated purpose of the questionnaire. But many of the questions are so badly framed that the reader cannot give an intelligent or clearly defined answer. Some questions ask for reactions to given circumstances and offer three possible reactions from which the selection is to be made, ignoring the fact that there may perhaps be ten or more possible reactions in real life.

This is all good fun and ideal for parties, provided it does not give readers the impression that survey questionnaires are based upon the same type of structure. At the same time it is the responsibility of those who frame statistical surveys to ensure that their questionnaires are free from the criticism which may justly be levelled at the others.

19 A Matter of Opinion

Opinion polls form part of the general structure of market-research procedure and as such they form a valuable part of that function. All that has been said about surveys in general applies equally to opinion polls in particular. In addition, the nature of these polls is such as to bring yet more difficulties in their path. When interpreting the results of opinion polls, it is first necessary to remember that one is dealing with opinions about facts rather than with the facts themselves. If a respondent states that he always uses a particular brand of some consumer product, that statement can be checked as a fact – if the interview is held in his home, for example, he can be asked to produce a specimen of the product or a wrapper or carton. But when he is asked what he intends to do at some time in the future, even if it is only a few hours ahead, his answer may be factual in that he is giving his true opinion but there is no guarantee that he will actually fulfil his expectations.

An opinion cannot always be lightly given. In giving an opinion, a person is revealing part of his innermost self; one need only read dramatic critics to appreciate this. To some members of the community it is a heaven-sent opportunity to air their views to someone who is being paid to listen; to others the effort of revealing what they really think demands a quality only just short of courage. In between these two extremes there are those, probably in the majority, who do hold definite opinions and are prepared to air them rationally, but there are also those who cannot really make up their minds. Many people, however, will not be able to give a considered opinion at a moment's notice. They need time in which to digest the question. The very essence of statistical interpretation – the time and the facility for studying problems from all angles before giving an opinion – is denied them. If the respondent were to be asked the same question an hour later or on the following day he might conceivably give a different answer. Opinion may veer like the wind, particularly on questions which appear trivial to the respondent.

The sponsor of a survey may be able to make profits from the knowledge of which one out of two advertising presentations makes the greater impact on the majority of respondents. To the latter, however, they are both advertisements and not very different at that. They cannot tell which makes the greater impact, for reaction to impact is an automatic function outside the conscious control and they are being asked to study the advertisements. Impact is a fleeting concept utterly broken by prolonged study. Nevertheless the respondent is asked for an opinion. If he is honest, he may say that he cannot see any difference; this will not help the sponsors except that, if everybody said the same, it obviously would not matter which type of advertisement was used. If, on the other hand, he gives a definite answer, it might not be of real value unless there is a very obvious difference between the advertisements, in which case the survey was probably hardly necessary at all.

Decision becomes even more complicated when there is not merely a choice between two possible answers. Does he like roast beef? The answer is *yes* or *no* and he can probably answer the question fairly easily. Does he prefer roast beef to roast lamb? This is not so easy. He may have no distinct preference; or at any specific moment of time, he may prefer the type of meat which is actually on the plate before him. When he is eating beef, there is nothing better. Even if he can say that he distinctly prefers one meat to another, the question as to why he has such a preference is more difficult. Is it the taste of the meat itself or of the sauces which accompany it; or is it the comparative tenderness or succulence of the meat which attracts him; is it because one meat is normally fatter than the other or is it because he usually receives larger helpings of it? Who can tell what visions he is conjuring up? All he may be able to offer by way of explanation, unless he is prompted, is that he prefers roast beef because he likes it better!

There are many questions the answers to which may involve

many shades of opinion. The respondent may find it difficult to classify these. This has always been so, as the Latin poet Terence noted: *Quot homines, tot sententiae.*[1] All subjects of choice have many facets. A person may prefer one facet for one reason while preferring another facet for some other purpose. In a survey of the popularity of different colours for ladies' handbags, some 75% of respondents liked red best, the criterion used apparently being one of mere preference between different colours.[2] However, it was also discovered that, notwithstanding this absolute colour preference, there was an overwhelming preference for black handbags when it came to actual selection for buying. The criterion in these circumstances was one of usefulness of the bag as a fashion accessory. A red handbag could be used only with matching or complementary colours of wearing apparel, whereas a black bag could be used with almost any dress outfit.

A preference today for something not possessed may well disappear once that thing is obtained, for anticipation is often much more acutely felt than is the realization. A man will desire something very badly until he has it, whereupon he will want something else. A child faced with the problem of choosing which of two Christmas presents he would like may reply with honest childish logic that he would prefer both! This is not mere greed; he may just not be able to make up his mind.

A further difficulty arises from the fact that an individual who is asked for his opinion will thereupon realize that he ought to have an opinion even if he does not really have one. This has become clear from experience in the use of television. An interviewer stops people in the street to seek their opinions on some matter of topical import, and it is often quite evident that some of those interviewed have never previously thought seriously about the matter at all. While it is true that television interviews have little in common with sample-survey interviews, it is also true that the people interviewed in each are drawn from the same populations.

Surveys of opinion clearly need very strict and tight control. All

1. 'So many men, so many opinions'.

2. G. Johns, 'Contributions from Some Personality Theories', *Commentary* (Market Research Society) No. 3. 1960.

this is well known to research technicians and every possible precaution is taken to overcome the difficulties. The dangers of taking opinions without such checks may be illustrated by the example provided by a survey of the shaving efficiency of razor blades as measured by the users' satisfaction. The majority of users to whom the new blades were issued reported that they had had an excellent shave on the first use of each blade but that the shaving efficiency of the blade had diminished with each use until, at a certain stage, the blade had no longer been capable of giving a good shave at all. The used blades were then returned.

A similar experiment was then repeated and again much the same results were reported – a good shave at the first use and a progressive deterioration of efficiency with each successive shave. Did these experiments prove that this is really what happens to the efficiency of a blade? Indeed, they did not; for the second experiment was carried out with the used blades from the first experiment. Nothing had been done to the blades, except to repack them into new packets, so that the first shave in the second experiment did not in fact represent the blade's first use.

What, then, did the experiments demonstrate? They revealed very little about the razor blades, but there are some other interesting pointers which repay investigation. The results proved nothing one way or the other about the actual shaving efficiency of the blade, although a question which might be further investigated is whether in fact blades, which lose efficiency with daily use, might recover their efficiency after a period of non-usage. If this theory could be proved to be untenable, then clearly the experimenters were merely recording their opinions of efficiency rates rather than recording the actual degrees of efficiency. Their judgements must have been influenced by the knowledge of their participation in the experiment. They expected to obtain better results with what they thought were new blades because that was the most reasonable expectation; and they managed to convince themselves that they had indeed achieved this expectation.

This particular point also involves the question of what is an efficient shave. The experiments may have provided some clues as to the criteria used by the sample members, and there can be little doubt that the members employed a number of differing criteria.

The satisfaction level of shaving depends not only on the blade itself; it depends also on the other obvious factor of what shaving preparation is used. It also depends upon the possibility that an individual's satisfaction level is itself an uncontrollable variable. This may fluctuate from day to day according to the individual's general health, as also it may depend upon whether he always holds the razor at the same angle, whether his skin is more tender or his hair tougher on one day than it is on another, or whether he sets his standards higher or lower according to the requirements of his social activities.

Another interesting control experiment was carried out by the Consumers' Association to establish what results were achieved by the use of hormone skin creams.[1] Two hundred women over the age of thirty-five were invited to assist in the experiment, one hundred being in each of two main groups A and B. Fifty members in each main group were given the specific hormone creams which had been repacked into plain jars; the other fifty in each group were given a specially prepared control cream which was similar to the hormone cream in consistency, colour, and perfume. The sample members were given no information other than that they were testing a hormone cream. They were given instructions as to the method of applying the cream and were asked to report progress after three weeks and also after six weeks applications. The results are tabulated in figure 38.

The main point of interest in this experiment was the fact that the control cream contained no hormones, yet the results it achieved in registering satisfaction were not far short of those achieved by the hormone cream. In Group B at the three-weeks stage it actually showed better results. A comparison between the three-weeks results and the six-weeks results reveals that, in each group, those actually using the hormone cream began to change their opinions. This numerical change may have had some meaning but there is not sufficient evidence to support this. The samples were relatively small. But why was the control cream, which contained no special ingredients, considered by so many women to be so much better than ordinary creams? The

1. Report published in *Which?*, February 1960.

Consumers' Association, while admitting the possibility that everyone's skin might have been improved during the test period because of the very fine summer or by the regular daily massage, concluded that the most likely explanation was a measure of the

Results reported	*Group A*		*Group B*	
	Hormone cream	*Control cream*	*Hormone cream*	*Control cream*
(*i*) *After 3 weeks:*				
No better than or not as good as ordinary creams	53%	57%	53%	36%
Better than ordinary creams	47%	43%	47%	64%
Total Number of Replies (Possible: 50)	47	49	50	47
(*ii*) *After 6 weeks:*				
No better than or not as good as ordinary creams	42%	57%	38%	42%
Better than ordinary creams	58%	43%	62%	58%
Total Number of Replies (Possible: 50)	42	45	50	42

Fig. 38

Skin cream tests

psychological effect produced by describing the cream as a hormone cream. The members examined their skin every day for traces of improvement, and many of them were able to 'find' what they hoped to find.

Whether or not either cream had effects differing appreciably from those obtainable from any standard cream is still in doubt in the absence of any independent measurement mechanism. It is possible, however, that both creams used in the experiment may have had beneficial results compared with the creams which the sample members had previously been using. It is important to note that the hormone cream was not being tested against the control cream. Each cream was in fact being judged against the variable standard of normal creams used before the experiment. There are many different 'normal' creams available. Members who reported progress with the experimental creams may previously have used inferior creams, whereas those who reported no progress may have been normally accustomed to a cream of similar quality.

The results of the survey were therefore inconclusive in testing whether hormone cream was better than others. The long-term possibility of the superiority of hormone creams, while admitted, was nevertheless discounted since treatment of this nature was advertised as having rapid results. The experiments may not have been very helpful to consumers, for whom the survey was originally conducted, but a manufacturer could possibly have found use for the type of results achieved if the samples had been larger. A manufacturer would be gratified that 50% of all women could find satisfaction in his products. Whether the satisfaction was real or merely psychological would be of secondary interest to him if he could be assured that there was a substantial potential market for his products. Again, the results achieved with the control cream were such that some lively manufacturer, noting the cream's remarkably simple composition and low cost, may already be using a similar formula.

This experiment holds pointers to other factors beyond the limited sphere of cosmetics. The sample members were asked to give a positive or a negative answer. Where actual results tend to be at the margin of observation so that people may have trouble in making up their minds, or where perhaps there is only an imaginary difference to be observed, may it not be that the probability of receiving a positive answer is likely to be about equal to the probability of receiving a negative one?

The opinions so far mentioned were expressed on matters which themselves were not of vital importance to the sample members.

Requests for opinions or for forecasts of future action on more personal matters may also be countered by a refusal or by a deliberately false answer engendered by a respondent's resentment to the question. Politics is just such a subject. Prior to the 1959 general election in Britain there was a great deal of nonsense talked and printed about opinion polls which had been canvassed at frequent intervals in order to ascertain trends in the electorate's feeling towards the various political parties. Much more was read into the results of these surveys than was ever there, and even more was written about the mythical 'don't-know' sect.

It is doubtful if there really are a great many people who 'don't know' how they will vote if asked one or two days before a general election, although the surveys continued to show a high response rate in this category right up to the day immediately before the election. It is highly probable that the 'don't-knows' included some few who really did not know but they included a far greater number of 'couldn't care', and 'won't vote', or 'mind your own business' and similar sentiments. There are a number of people who refuse to vote because they do not believe that the ballot is really secret. These are extremists, perhaps, but secrecy as to one's voting activities is accepted by very many people as an essential part of their participation in the ballot. To be asked before the event what they are going to do later in secret destroys the whole idea of secrecy, and it is not surprising if such people consider that questions of this kind constitute unjustifiable prying into their affairs. This resentment may not always take the form of refusing to answer questions but it may lead to some deliberately misleading replies.

Another point of interest centres on the percentage of those interviewed in surveys who express an opinion as to how they will vote. This percentage is usually higher than the percentage of the total population actually recording votes. This can only mean either that the sample interviewed was not truly representative of the total population or that the responses indicated expectations which were not translated into actions. In the weeks preceding an election, results of new surveys are published almost daily. These results naturally vary as between themselves and there is a tendency to consider each of these as an accurate reflection of the state of the political parties as at the relative dates. The implication is that the results may be charted so as to define a time-series chart representing changes in the state of the poll. In truth, however, this is absurd, as also is the talk about the 'narrowing gap' between the respective parties' popularity levels.

The daily changes are so small that the relation between the different observations cannot be regarded as meaningful. More important still, however, is the apparent ignorance of the fact that different surveys obtain different results merely because they record the opinions of different people. The samples are not identical and consequently the sample statistics will differ also. It will be remembered that sample statistics are only estimators of the true value of the population parameter. All the sample statistics are subject to sampling error. All of them may fall within a certain specified range about the population parameter but it is unlikely that any one of them will be exactly equal to the latter.

If the same sample is used throughout a number of surveys, then the variability of a party's percentage popularity within that sample may legitimately be charted to record the changes noted over a period of time. Where different samples are used, however, the values derived from each of them cannot be plotted and linked together on a chart in the same way. To chart the values in this way would be parallel to plotting temperature readings on consecutive days to show relative daily changes, only to discover that the first day's observation was taken in London, the second in Paris, the third in New York, the fourth in a boiler house, and the fifth in a refrigerator.

Sample statistics are only approximate values for the true population parameter, although some results are published to a fine degree of apparent precision. The use of such spurious accuracy is clearly unjustified in the translation of survey results into forecasts of actual results, even if such a translation should ever be attempted at all. In a voting population of some 35½ million, a difference of only one per cent is equal to about 355,000 votes. One party's gain may be another party's loss, unless the voters stay away from the poll altogether, and the effect of each gain or loss is in fact doubled in respect of the total votes cast for the respective parties. The effect of such a variation in marginal constituencies would be more than enough to change the whole structure of election results in terms of seats won.

In this respect political-opinion polls differ quite definitely from ordinary surveys. A manufacturer would not be unduly perturbed if the results of surveys indicated that his share of the market varied from 48% to 51%, instead of remaining level at 50%. In an election survey, however, such approximations will not suffice for successful forecasts. In the marginal constituencies, results are decided by exceedingly narrow margins; so narrow that they would defy recognition in pre-election surveys. The polls promoters were able in 1959 to point to the main fact that all the published opinion polls had pointed to a Conservative victory on a minority vote, although they failed to gauge the extent of the margin of victory and also failed to detect that there were two opposite swings of opinion – towards the Conservatives in the south and away from them in the north. With such a large margin of victory as actually occurred, the failure to estimate the extent of the overall nett swing towards the eventual victors was not sufficient to invalidate the main conclusion. In a closer contest, however, sampling errors could easily have led to a false conclusion.

In this connexion it is worth while to note that as soon as opinion had been translated into fact – that is, once the voting had actually been completed – it was possible after the proclamation of about fifteen results to forecast the final result with remarkable accuracy. Thirteen minutes after the first result had been announced Reuters, who had the use of an electronic computer, were able to report that the early results showed that a Conservative victory with an increased majority was a virtual certainty. The computer analysed results as they were received and compared them with the corresponding results in the previous election, thus discerning the percentage swing in individual constituencies. Half an hour later the computer was forecasting a majority of 90, and a further half-hour later it was predicting a majority of 105. The actual majority was in fact 100. These forecasts were possible because the statisticians no longer had to rely on opinion. With real facts in their possession and with the necessary mechanism to

ceed with confidence to

the promoters have to
olls are conducted as
not all the difficulties
the superficial devices
n for thinking them to

trends might indicate at what point the main election campaign should be mounted.

Opinion polls are of much less usefulness to the ordinary rank and file of voters, and their interest in the apparent trends revealed is also a doubtful factor. Yet, in the absence of evidence to the contrary, the poll promoters might reasonably claim to have been prime movers in the unexpectedly high level of stock-exchange activity which occurred immediately before the election. If the investors really believed in a Conservative victory, on what were they basing their confidence other than on the poll results? It might be, of course, that the ordinary person was able to scent the result without any help from surveys. Indeed, perhaps the share index is as good a pointer as any to an election result. People's actions as reflected by the buying or selling of shares, and as translated by the brokers, may perhaps speak more loudly than do answers to a questionnaire, although the population of shareholders of course forms only part of the total population of voters.

It would be interesting to know what influence, if any, the publication of opinion poll results has upon the way in which individuals actually vote. The publication of some statistics may itself affect the state of affairs which the statistics record, in the same way as television cameras, positioned to photograph crowd scenes, will themselves ensure the presence of a crowd to watch the cameras photographing the crowd. Do the polls merely reflect opinion or do they themselves have an effect on the opinions recorded in subsequent surveys? This question is not unlike that which seeks to decide whether newspapers reflect national opinion or whether they mould it. The answers to the questions, however, are not necessarily parallel. Does the voter react individually to the 'facts' revealed in the survey? Does he, because of the realization that his party is losing ground, thereupon resolve to vote when otherwise he might not have done so? Or does an apparent gain in the popularity of his party induce a mood of indifference – that they will be all right without his vote? The Liberal Party has indeed claimed that a Liberal candidate lost an election to his Conservative opponent because people who would otherwise have voted Liberal in fact voted Conservative in face of the fear that, as the opinion polls purported to show, Labour appeared to be gaining strongly.[1] The Liberal Party actually passed a resolution recommending that the publication of poll results should be prohibited for four weeks before election day. If, indeed, such reactions are real, it is clear that no pre-election survey could ever hope to produce the same pattern as the actual election results.

Perhaps the main value of election surveys is to the promoters themselves and thereby to the improvement of research standards generally. The results of the surveys, although not intended to be strictly comparable with election results for the purpose of forecasting those results, can nevertheless be compared with them from a somewhat different aspect. The object of research is to ferret out statistics not otherwise available; if they were readily available, research would be unnecessary. Election surveys are different in that the statistics they produce purport to reveal a state of affairs which, in the short term, fluctuates within fairly close limits. The survey statistics and the actual results should therefore both reflect approximately similar states. They may thus be compared and the degree of relationship discerned may be used to test the efficiency of the application of sampling and interviewing techniques employed in the survey. The closer the comparison then, by inference, the more efficient the methods, and, therefore, the greater the likelihood that they may enjoy the same accuracy in other fields of research where actual dimensions cannot be measured.

1. *Daily Telegraph*, 27 May 1960.

20 Under Control

One of the most important applications of statistical theory is to be found in statistical control techniques, of which quality control is a well-known example. This is often thought of as being purely an industrial discipline because it is in industry that it has had its widest adoption. The control theories and techniques may, however, be just as usefully applied to many processes of a repetitive nature in almost every other sphere of activity.

Since the risk of error in sampling methods is larger when small samples are used, the sample statistics will fall within a comparatively wide range and minor fluctuations are likely to be concealed. On the other hand, small samples nevertheless stand a reasonable chance of revealing sudden major changes. If a specified margin of error is acceptable, then it follows that minor fluctuations within the marginal limits may be ignored, provided that a major fluctuation, which provides an observation beyond either of the limits, may be instantly recorded. Such is the basis upon which the theory of control is based.

The technique of quality control has two main applications. The first application is in *process control* in which an actual process, such as the performance of a machine, is measured to assess the present performance and by implication to provide a guide to short-term future performance. The second application is in *acceptance inspection* which evaluates past performance of the process by measuring the quality of the goods produced. The latter application therefore deals with a finite population of things that have already been produced, whereas process control aims at checking the process during the course of actual manufacture. This enables the management to discern a defect in a process almost as soon as it occurs and so to prevent the production of faulty goods arising from that defect. Thus, while acceptance inspection deals with the finding of bad fish in a consignment, process control seeks to eliminate the diseased fish further up the river of production.

Both techniques depend upon the drawing of small samples from the respective populations and, from the quality measurements derived from these samples, making decisions affecting the whole population. Quality control is perhaps unfortunately named, for the word *quality* has so many different meanings or shades of meaning, but it is difficult to think of a better name. Quality is generally thought of as an abstract idea, and abstractions do not lend themselves to measurement. In quality control, however, the word *quality* has a restricted practical meaning for each of the control systems in which the technique is used. Quality is in fact defined in terms of measurement of the object or process measured. A product is produced for a specific purpose and an ideal process would produce perfect products. That is to say that the length, width, shape, weight, colour, and every other attribute of every product from that process would be identically perfect. Perfection is also difficult to define but in this connexion it means that the process should work exactly according to plan; and the quality of a particular attribute of the product of that process is measured in terms of its failure to satisfy the demand for perfection. This is a negative but logical method of measurement. We cannot define quality in such a way as to measure it positively other than to treat it as being synonymous with perfection, and the quality of an attribute is best measured in relation to its closeness to perfection or by the degree of its departure from perfection.

Mankind has not yet been able to construct any process which will give consistent results continuously – nor is it ever likely to do so. Machine parts will wear out or come out of alignment, while human error may account for many other faults. It is regrettable that we have to admit that perfection exists only in theory; nevertheless we have to accept a qualified perfection somewhere within tolerances permitted by the specifications for the work to be carried out. If a specification measurement is one inch, plus or minus five-thousandths of an inch then any measurement between

0·995 and 1·005 inches will satisfy this qualified perfection and the quality of the measurement will be acceptable. We may, therefore, ungrammatically re-define quality as qualified perfection.

The object of process control is to ensure that the specified statistical tolerances are being held and to detect faults as soon as they occur. It allows for variability within the permitted limits, and any two or more products whose measurements fall within these limits are treated as being identical in the sense that they conform to the requirements. This is so, even although they may have slight differences between themselves, in the same way as so-called identical twins also have very slight differences. Variability itself is not significant, but the degree of variability and its direction are both significant. The extent of variability present in a process will be revealed by the range of measurements, and a wide range of measurements will point to some process defect as the cause of the fluctuations. On the other hand, a trend chart will show up any significant shifting away from the accepted standard; consistently high or low measurements will ordinarily point to the faulty setting of a machine rather than to any mechanical defect.

A process is said to be in statistical control or is described as stable when the variability in the recorded measurements is such that it may be attributed to chance inasmuch as it could occur in random sampling from an homogeneous population. The process is therefore not actually controlled by statistics but, instead, statistics checks the performance to ensure that the process is not out of control. If such variability as occurs could be attributable to chance, then the process is in control. If in such circumstances the results achieved are not satisfactory to meet the specifications, then it is the process itself and not its minute-to-minute performance which is at fault. Such a state of affairs exists when the setting of the process is faulty, and the results obtained will indicate that the process requires to be re-set. When, however, the variability cannot be attributed entirely to chance, the process is out of control and it is this type of problem with which quality control is largely concerned.

This may seem to place too heavy an emphasis upon fault-finding. Quality control is naturally just as concerned with the correct functioning of a process but it is only when faults appear that action becomes necessary. Nevertheless it is desirable to dispel any impression that quality control is merely a fault-finding device. Much of its responsibility is to hold a watching brief over the process. It is always the freak case which is reported in newspapers; everyday humdrum affairs are rarely mentioned yet they form the standard against which the freak is distinguishable. So it is in quality control; when a process is in control, it is better to leave well alone unless it is clear that the control may be tightened still further.

Control techniques are based upon the properties of the normal curve. It has been shown[1] that about 99·7% of all observed values drawn from a normally distributed population are contained within the interval of 3 standard deviations on either side of the mean value and that, therefore, only about three in every thousand of such observations will fall outside those limits. A control chart may therefore be drawn as in figure 39, showing possible values on the vertical axis and a series of consecutive integers, representing the sequence of observations, along the horizontal axis. A horizontal line is drawn at the height corresponding to the mean value; and horizontal lines are also drawn at heights representing the control limits. The upper control limit is drawn at a height corresponding to the value of the mean plus three standard deviations; the lower control limit is drawn at a height corresponding to the value of the mean minus three standard deviations so that some 99·7% of all observations should fall within those limits.

It is usual to use control limits at 3 standard-deviation intervals. This is not essential but is generally adopted because it provides more conservative limits than would limits placed at, say, 2 standard-deviation intervals. If the permitted deviation range were to be narrowed in this way many more values would fall outside the control limits. Some 32% would in fact fall outside the limits and this would rob the control system of its desired measure of discrimination. On the other hand, if the standard-deviation range is widened to say, 6 standard deviations, the freak values might still fall within the control limits and thus pass unnoticed.

1. Chapter 15.

A sieve must be just so fine that it will let fine earth pass through but will retain the larger stones and pebbles. Control limits must be so placed that they allow the minor fluctuations to pass while at the same time drawing our attention to the major fluctuations which may mean real trouble.

The mean value and the standard deviation required for this

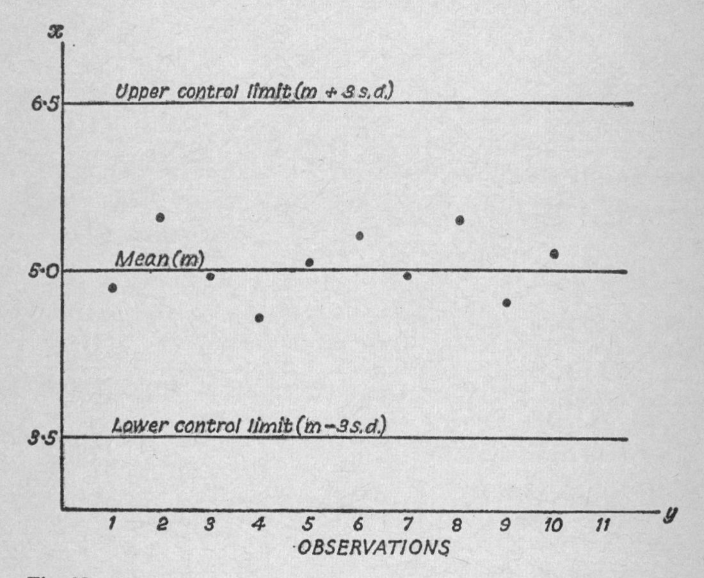

Fig. 39

Control chart (mean = 5·0*; standard deviation* = 0·5)

chart are strictly population parameters. It is in the nature of things that such parameters are rarely known and, instead, it is necessary in practice to derive estimates of these from samples taken in the past and to revise these estimates as more data become available. The individual observations are plotted by dots and it is simplicity itself to see whether or not they are within the statistical limits defined.

Figure 39 gives a representation of the basic idea, but in

practice it is usual to plot the arithmetic means of small samples, the sample size very often being no larger than 4. There are a number of good reasons for this. Firstly, the individual observations may not be normally distributed; on the other hand, small samples are much more likely to be sufficiently nearly normal. Secondly, the individual observations may fluctuate widely within the accepted limits and it is necessary for the control limits to be set fairly wide apart if the scale of the charts is to be sufficient to show up the differences. Where the limits are far apart, it is not always easy to be able to discern significant changes as soon as they occur; the chart is then apt to develop a measle-like appearance with masses of dots obscuring the overall pattern. The variability between sample means, however, is much less than the variability between individual observations. Thus, if only the means of samples are plotted, the control limits may be set closer together and it is much simpler to discern significant changes in the pattern.

It is true, of course, that the plotting of sample means may perhaps involve the missing of an occasional value which falls outside the control limits. The presence of a freak without-control value (that is, for example, one of the three out of one thousand that may be expected to fall outside the limits) may escape unnoticed because its effect may be largely negatived by the other values in the sample. This is not of itself important. These freak values are accepted risks. Where, however, such values recur persistently so that their appearance cannot be due entirely to chance, it is important that they should be recognizable and this is more rapidly possible with charts which plot sample means.

Sample means are used specifically so that the variability of individual values shall be damped down, but this variability cannot be ignored and separate charts are therefore used to record the ranges of values in the samples. The necessity for control of range values, in addition to mean values, arises from the fact that two samples may have identical mean values and yet have quite different range values. It is indeed possible that a mean value may be within the control limits whereas all the individual values may be outside the limits. Such a sample, consisting of four values, might be as follows, where the lower control limit is 0·995 inches and the upper limit is 1·005 inches.

First value	0·990
Second „	1·007
Third „	1·010
Fourth „	0·993
Sample total	4·000
Sample mean	1·000

This mean value coincides with the assumed population mean from which the limits were measured. It gives a misleading appearance of perfection, for each of the individual values is outside the limits. This possibility arises because of the wide range of variability within the sample. The process is fluctuating between high and low values and is no longer stable.

The main advantages to be derived from process-control techniques arise from the fact that, although they dispense with the necessity for one hundred per cent inspection, they can nevertheless provide methods of controlled inspection at any stage of production. By revealing the movement of a process out of control almost as soon as this occurs, they reduce the wastage of labour and material which would otherwise have been lost in the production of faulty products. The type of chart employed is basically simple, the principles are easy to grasp and can be understood without reference to involved mathematical calculations. But its superficial simplicity may at times be misleading since the assumptions involved (e.g. the normality of the distribution) may not in fact represent the true state of affairs. It should also be noted that the technique is designed to show when a process is no longer in control and that, although it may point to a possible general type of fault, it goes no way to showing what amount of correction is required to bring the process back into control.

In addition to keeping watch for the freak observations, it is also important to observe the length of run of observations between the appearances of the freaks. A run of observations will reveal any shifting of the process mean away from the assumed population mean even though all of the recorded observations still fall within the control limits. The fact that a process is in control is not sufficient justification for not altering the process in such a way that the control limits could be narrowed. To develop the usefulness of this application, use is sometimes made of cumulative charts on which are recorded, not the individual values, but the cumulative totals of values. These charts show relevant information remarkably clearly. The value of each individual measurement is represented by the slope between the two consecutive points, and a line drawn through all the points will give a directional picture of the process as a whole. Any real change in direction (i.e. of steadily increasing or decreasing measurements) will at once be apparent.

There is one technical difficulty about the drawing of such a chart; the line drawn is likely to run rapidly across a sheet of squared paper, possibly leaving the upper part of the sheet untouched. This can result in a sizeable wastage of paper. One way of reducing this wastage is to wrap the paper around a long cylinder so that the line will not reach an end of the paper until it reaches the top of the cylinder. The line will in fact describe a spiral movement around the cylinder and, unless its slope is steep, it will make far better use of the paper. The cylindrical presentation also helps to make the reading of a line much simpler. The line's direction may easily be discerned by turning the cylinder about its axis. If the charts are drawn on flat paper there will be many continuation sheets required; the more sheets there are, so the less easy it is to obtain a rapid and clear notion of the data presented. Cylindrical presentation also helps to save space since it avoids the necessity for the spreading out of the continuation sheets which would otherwise be necessary.

Process control is a series of techniques for checking a process during the course of its performance. The other main form of statistical control, acceptance sampling, is concerned with the sampling inspection of finished products to determine whether the batches from which the samples are drawn are of acceptable quality. Here it should be noted that finished products are not necessarily in the finished or assembled state in which goods are purchased by consumers. Any product which has been subjected to a process is, for our purposes, a finished product with regard to that process. Acceptance sampling thus occurs after a process has been completed. If a batch of goods has been produced at an acceptable quality level by a process which has been subject to process-control observations then, so far as the manufacturer is concerned, acceptance sampling becomes unnecessary since its work has already been done at the production stage. The buyer of such goods, however, may not have access to the process-control information and will then have to devise his own acceptance-sampling scheme. In this connexion it is interesting to note the development, notably in America, of the insistence by some large firms that each batch of goods supplied to them shall be accompanied by copies of the manufacturer's process-control charts.

Both forms of control are based upon evidence derived from random samples and both employ the basic idea of control limits, but there is nevertheless a fundamental difference between them. When process-control techniques reveal that a process is not functioning correctly, the process may be adjusted almost at once. The emergence of a few defective items will therefore lead to a correction of the process and thus also to the cessation of production of defective items. Acceptance sampling, however, is concerned with a finite population of products which have already been produced. This is its prime concern and, although the information gained from it can and does have an eventual bearing upon the process which produced the goods, it cannot alter the quality of the goods in the batch which is being inspected.

The probability of a batch being accepted depends upon the true quality of the goods in the batch. The probability of acceptance is the probability that, if a batch of a specified quality is offered, it will be accepted. The probability of acceptance is not the probability that if a batch is accepted it will be of a specified quality. Every acceptance-sampling scheme involves an element of risk and some batches with a defect percentage higher than the specified percentage will be accepted. The risk involved in a particular scheme for example might mean that it could occasionally accept batches containing 20% defective items. This would not mean that any of the batches were necessarily 20% defective. The level of quality of goods in a batch exists quite independently of the ability of inspection schemes to measure that level. Inspection can have no effect on the actual quality of the goods; it can only attempt to measure the level of that quality.

An idealized form of sampling scheme would reject every non-acceptable batch and would accept only acceptable batches. Such perfection would, however, require almost the equivalent of 100% inspection, thus defeating the objects of sampling. Suppose that the acceptable quality for a particular product is represented by there being not more than two items defective in every hundred. The maximum percentage defective allowable is thus 2%. Suppose also that the batch size is 20,000, that the huge sample of 19,999 is drawn from the batch and that 400 defective items were discovered in that sample. Despite the size of the sample inspected, we would still not be sure that the batch was not more than 2% defective. The item excluded from the sample might, on inspection, be found to be defective, so that there would then be 401 defectives in the batch of 20,000. This would represent a defective rate of just over 2%. This explains why some latitude must be permitted in the region of 2% defective and also illustrates the nature of the risks involved in acceptance-sampling schemes. Certainty in acceptance schemes is not possible.

The mathematical bases upon which acceptance schemes may be constructed for their various purposes are too involved to repeat here. We may, however, note the forms which such schemes might assume. A simple acceptance scheme which accepts a quality of 0·5% or fewer defectives in a batch would be based upon the drawing of a sample of 75 items. If none or only one of the items is defective, the batch is accepted but, if there are two or more defectives, the batch is rejected. Rejection in this context may mean the return of the whole batch to a supplier or it may merely be the rejection of the sample followed by a 100% inspection of the batch. The second definition is the true statistical one. The physical rejection of a whole batch may result from the rejection of a sample but it is not a necessary outcome of the latter.

Alternatively a double sampling method may be employed.

For the same acceptable quality level (i.e. 0·5% defective) a sample of 50 items might be drawn. If none is defective, the batch is accepted. If three or more items are defective, the batch is rejected. If, however, only one or two items are defective, the double sampling method develops by selecting another sample, this time of 100 items. Then the results of the two samples together will decide upon acceptance or otherwise. If there are two or less defectives in the total of 150 items, the whole batch is accepted; whereas the batch is rejected if there are three or more defectives in the total of 150 items in the two samples. Multiple sampling schemes develop the idea behind double schemes – the reduction in sample size and the subsequent combination of results from a number of samples if this should prove necessary.

Sequential sampling in effect develops the idea of multiple sampling to the limit of its application. It differs, however, in that the sample taken is built up one item at a time or, in other words, that samples of one item are taken successively and the cumulative result from all samples is then considered as if it were derived from one single sample consisting of all the separate items in the separate samples. Sequential sampling has the advantage that a decision may be made with a smaller average sample size than is required by other methods. It is therefore especially valuable where inspection is costly or where it involves destruction of the object tested.

Development of acceptance-sampling theory has also produced deferred sentencing schemes in which the acceptance or rejection of a batch may partly depend upon the results obtained from previous and subsequent 'neighbouring' batches as well as upon the results derived within the batch itself. The advantage of this system is that by combining results it can give a better discrimination against the acceptance of bad quality, if the quality level is reasonably constant, than can a multiple or double scheme using the same sample sizes confined within separate batches. Alternatively it can achieve the same results as the latter methods but with smaller sample sizes. Where the quality level is variable, however, the discrimination offered by these deferred sentencing schemes is not so good, nor is it as sensitive as ordinary schemes to a change of quality level. They also suffer from the fact that batches are likely to be accepted or rejected together and this may interfere with practical production considerations by upsetting the flow of work.

In Chapter 15 it was noted[1] that the probabilities of 0, 1, 2, or 3 defectives respectively in a sample of 3 are represented by the successive terms of the expansion of $(q + p)^3$. If the quality level of a batch is 10% defective then the probability of a defective occurring is $p = 0{\cdot}1$. The respective probabilities are therefore

No defectives	Probability $= 0{\cdot}729$
1 ,,	,, $= 0{\cdot}243$
1 ,,	,, $= 0{\cdot}027$
3 ,,	,, $= 0{\cdot}001$

These may be represented by a histogram (figure 40) in which the probabilities are expressed as percentages of samples which over

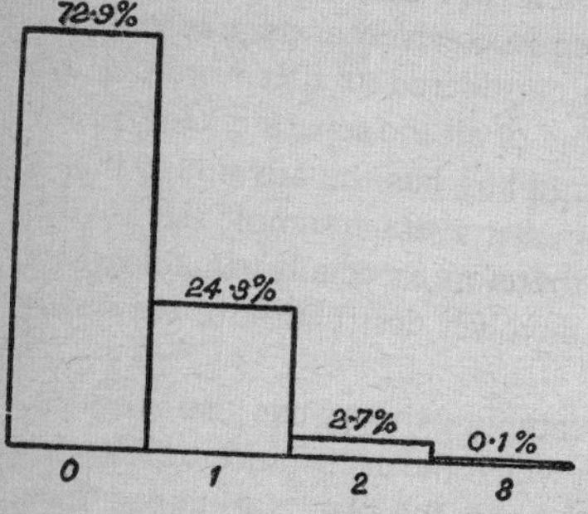

Number of defectives in samples of size 3.

Fig. 40

Percentage of samples with the stated number of defectives (Binomial distribution; $p = 0{\cdot}1$)

a long run would be expected to reveal the respective number of defectives.

The nett result therefore is that nearly 73% of all samples will give the impression that the batch from which they are drawn is

1. Page 219.

perfect since they show no defectives at all. 24·3% of the samples will suggest that the batch is $33\frac{1}{3}$% defective (i.e. one defective in every three items). Under 3% of samples will suggest that the batch is $66\frac{2}{3}$% defective, while 0·1% of samples will give the impression that the batch is 100% defective. There is a paradox to be noted here. The known value of p is 0·1, that is 10% defective, yet nearly three out of four samples (actually 72·9%) give the impression that the batch is perfect and not one sample can give a clue to the real value of p.

Sample size is an important consideration and, indeed, as has been noted elsewhere, the reliability of a sample depends primarily upon sample size and not upon the population size. Population size has very little effect. If, in a sampling scheme where a batch is accepted only if both items in a sample of 2 are non-defective and the batch quality level is 10% defective, then in a batch size of 20, there would be 2 defectives. The probability of a random sample being acceptable would then, by the multiplication law, be equivalent to

$$\frac{18}{20} \times \frac{17}{19} = 0{\cdot}8053$$

for a batch size of 100, the probability of acceptance would be

$$\frac{90}{100} \times \frac{89}{99} = 0{\cdot}8091$$

while for a batch size of 1,000, it would be

$$\frac{900}{1{,}000} \times \frac{899}{999} = 0{\cdot}8099$$

These probabilities do not differ very greatly and larger batch sizes produce probabilities which approach the limiting value of $0{\cdot}9 \times 0{\cdot}9 = 0{\cdot}8100$.

The size of the batch is therefore of little consequence, but nevertheless it is of *some* consequence and it may well be that in certain circumstances it might be advantageous to seek the greater discrimination available in an increased sample size. This may be so where the batch is sufficiently large so that the increased cost of taking larger samples may be spread over a large number of produced units.

On this evidence, small samples would seem to be of no great account, yet they are used extensively in quality control. How is this? The answer really is that the samples do not exist in isolation. Although a sample size of four may be used, samples are taken regularly and it is the effect of a number of samples considered together which really gives the call to action. Individual violations of the control limits are not by themselves sufficient to show that a process is out of control. Some violations are to be expected in conformity with the calculated probabilities of their occurrences, but when they occur we must be able to recognize whether they are the chance freaks or the progenitors of a race of deformed products. The first action called for is to draw another sample. If more apparent freaks appear then the process will clearly require some detailed investigation to discover the cause of the trouble.

The theoretical justification for the adoption of statistical control techniques, based on evidence derived from samples, rests upon mathematical probability theory. Where absolute certainty is required the techniques are not suitable for they are not designed to deal with certainty. But if near-certainty will suffice, then they may be very valuable. Theoretically the only way in which to achieve certainty is to carry out 100% inspection yet, in comparing control schemes with full inspection, one must accept that in the long run inspection is itself imperfect. Inspection theory may be unassailable but the inspectors are not so reliable. The greater the number of items inspected so the less likely is a solitary defective item to be discovered. It is a paradox that the higher the quality of a product so, relatively speaking, the poorer is the standard of inspection. Where there are very few defectives to be found, so many perfect ones will be inspected and passed that the inspection may tend to become automatic. The inspectors will form their own intuitive ideas of the probability of discovering a defective and will regard this likelihood as remote. Since no defectives are expected, they may not be noticed – they all look alike after a while.

21 Linear Programming and Games

With the ever widening horizons of applied mathematics in economics, industry, and operational research generally, statistics is gradually absorbing many new and powerful branches of mathematics. Applied statistics as a subject is becoming more and more identified with the study of the problems with which statisticians are expected to deal, whether these problems may or may not be strictly classified as an integral part of statistical theory. The interpretation of statistics as a decision-making process, however, inevitably brings within its scope those mathematical techniques which assist in the process of making decisions.

There are a number of useful techniques on this mathematical margin. Linear programming is one of the relatively modern developments, and this is a powerful technique which, provided its basic assumptions are valid, can indicate a definite conclusion as to the best utilization of available resources in given circumstances. Classical statistical theory is concerned with inferences which may be drawn from a correct interpretation of data. Linear programming has quite a different function. Instead of assisting to make decisions by facilitating the drawing of inferences by sampling or correlation or any of the other standard statistical methods, linear programming provides a technique which in effect says '*this* is the right decision'.

It is the element of controlled watchfulness provided by statistical-control methods which is so invaluable but, as has already been noted, these methods do not themselves exercise any real control. Instead, they are the aids to navigation whereby management may steer through incompletely charted waters with the greatest assurance of safety. It is a tentative sort of control.

Linear programming provides a different form of control in the sense that management is thereby enabled to a large extent to have control of its own immediate fate. It is a technique which was evolved to solve what has become known as the 'transportation problem' and the more specific problem of how to achieve the optimum allocation of empty ships to ports where goods are awaiting shipment. The technique, however, has been developed in application to many economic and industrial problems which require the allocation of available resources to meet the varying demands which may be made upon them.

The principles involved are best demonstrated by reference to a particular problem. A manufacturer who produces three different products has three factories each of which is equipped to manufacture any of the products. No two factories are identical, however, and for a number of reasons the product profitability varies as between products and also as between factories. One factory, for instance, can manufacture product A more cheaply than it can manufacture product B, whereas in another factory this position is reversed. Again, one factory can manufacture some of the goods at a lower cost rate than another factory. The manufacturer receives orders for his products and thereupon wishes to plan the production throughout his factories in such a way that he may make the greatest possible amount of profit.

Orders are received for the following amounts:

Product A	50 units
" B	30 "
" C	20 "

The capacities of the individual factories and the profit per unit from each of the products manufactured in the respective factories are as in figure 41.

It will be seen that complications arise in that the relative productive capacities of the separate factories are different and also in that the total overall capacity of all factories combined actually exceeds the total of orders received. The manufacturer thus has some freedom to manoeuvre since not all his productive capacity will be taken up, and he will want to use only those production lines which together will provide the greatest amount of total profit. He will prefer to leave his least profitable production line idle. But his freedom is limited by the fact that, although factory No 1 can manufacture product A more profitably than any other factory, this particular factory has a productive capacity which falls short of the total demand for product A. Some units of this product must therefore be manufactured elsewhere.

There is therefore a form of competition, among the products ordered, for the available capacity of the respective factories. The

Factory Capacities and Profit

Factory	*Capacity units*	*Profit (£) per unit produced*		
		A	B	C
1	40	12	11	10
2	30	10	11	12
3	50	9	9	10

Fig. 41

problem is how to allocate the total itemized production among the different factories, a special form of the general problem of how to make the best allocation of the available resources to meet the demands made upon them. The complications involved in this type of problem make it impossible to attempt a solution by ordinary trial and error even where the number of the various demands and the number of supply sources are relatively small. Linear programming, while still depending upon the principle of trial and error, sets about the problem in a systematic way and, at the relevant stage, indicates that the best possible solution has been reached.

In the example there are three sets of data: (a) orders received for the individual products, (b) factory productive capacities, and (c) individual factory profitability rates for each product. These data form the raw material of a linear programming problem and may be summarized as in the grid in figure 42.

This is then expanded, as in figure 43, to relate factory profitability to factory capacity. This tabulation must take account of total productivity. As the total of orders is lower than the total productivity, we bring in a fictional product X in order to balance the totals. Product X should in fact be regarded as one which *could* be made but which is not made because it is not wanted. It thus represents wasted productivity. Its presence in the tabulation, however, is as important as utilized productivity since to

Products	*Profit (£) per unit produced*			*Total product units ordered*
	Factory 1	*Factory 2*	*Factory 3*	
A	12	10	9	50
B	11	11	9	30
C	10	12	10	20
			Total	100

Fig. 42

utilize the most profitable productivity is merely another aspect of not utilizing the least profitable productivity.

In figure 43, the number of pounds sterling shown in the larger squares are the relative profitability rates per unit product for each factory. The number in the top left-hand corner of each of these squares represents the number of product units allocated to the respective factory.

The allocation of factory capacities in this figure is purely an arbitrary one and represents a *possible* solution, such that all orders can be met by the production allocations without, at this stage, any regard to the profitability accruing from such an arrangement. This will rarely be the best possible deployment of resources, but it provides the basis for the ensuing calculations. It therefore does not matter how this first allocation is effected as long as the total productivity is allocated and all the orders can be met. This clarifies the reason for the introduction of the fictional product X; allocations of productivity for this 'product' represent

PRODUCT	FACTORY			Total product units required
	1	2	3	
A	40 £12	10 £10	0 £9	50
B	0 £11	20 £11	10 £9	30
C	0 £10	0 £12	20 £10	20
X	0 £0	0 £0	20 £0	20
Total product units allocated	40	30	50	120

Fig. 43

Factory profitability per product unit, and allocation of product units per factory

capacity not taken up. Such capacity cannot produce any profit and it therefore follows that the factory profitability in producing product X (i.e. of not producing at all) is nil, irrespective of which factory is involved.

This first possible allocation was made as follows. Fifty units of product A are required. Factory 1 can make only 40 units; these are all allocated to product A, leaving a balance of 10 units of product A to be allocated elsewhere. The latter are allocated to Factory 2, thus leaving a further 20 units in this factory to be allocated. Thirty units of Product B are required. Only 20 of these can be accommodated by Factory 2, and the balance of 10 units is allocated to Factory 3. Finally, 20 units of Product C are required and these are also allocated to Factory 3. Altogether, therefore, 30 units have now been allocated to Factory 3 (i.e. 10 of Product B and 20 of product C) thus leaving a balance of 20 units to be allocated to the fictional product X.

This, then, is just one possible arrangement and the total profitability is as follows

Factory 1	40 units A @ £12 =	£480
,, 2	10 ,, A @ £10 =	100
	20 ,, B @ £11 =	220
,, 3	10 ,, B @ £9 =	90
	20 ,, C @ £10 =	200
	20 ,, X @ £0 =	0
	Total	£1,090

Allocation of Product Units per Factory

Product	*Factory* 1	2	3	*Totals*
A	40	−1 10	+1	50
B		+1 20	−1 10	30
C			20	20
X			20	20
Totals	40	30	50	120

Fig. 44

The next step is to ascertain whether a more profitable overall allocation is possible. In the above example, for instance, units of A and B were allocated to Factory 2, whereas this factory has a higher profitability when producing Product C.

The method of finding the most profitable overall allocation is a process of elimination whereby, at each stage of our calculations, it is ascertained what additional profit could be obtained by taking one unit from an existing factory allocation square and placing it in an empty square. In figure 44, the first empty square is 3A (i.e. Factory 3 product A). We therefore place 1 unit in this square, but this involves the taking of 1 unit from some other square and this must be done in such a way that the totals of the allocation rows and columns remain unaltered. If, for example, we take 1 unit from square 3B, we add 1 unit to 2B and deduct 1 unit from 2A, as in figure 44.

The resulting change in profitability is

1 unit in 3A	plus £9
1 unit taken from 3B	minus £9
1 unit in 2B	plus £11
1 unit taken from 2A	minus £10
Nett change	plus £1

Thus, by moving 1 unit into 3A in this way, it is possible to increase the total profit by £1. Similar calculations are then carried out for every other empty square until it is found which of them will produce the greatest increase in profit, taking into account the necessary rearrangement of the existing allocation. No better allocation can be found at this stage and we therefore transfer as many units as possible into square 3A. The greatest number which can be transferred in this way is equivalent to the lesser of the totals in the square from which units are to be deducted. These squares are 3B and 2A, the original allocations of which were 10 units in each case. We therefore transfer 10 units to square 3A and make the necessary adjustments in the other squares. The revised allocation is then as in figure 45.

The total profit arising from this allocation would be £1,100. This, however, is not necessarily the best overall allocation and the process is repeated of testing whether the movement of one unit to another square would increase the profit. In point of fact, however, this profit cannot be increased in this instance and the present allocation is therefore a best solution. There are, in fact, other solutions which will produce the same amount of profit;

Revised Allocation of Product Units

Product	*Factory*			*Totals*
	1	2	3	
A	40		10	50
B		30		30
C			20	20
X			20	20
Totals	40	30	50	120

Fig. 45

any of these would be a 'best' solution. In practice it is not likely that a final solution would be obtained so quickly and it would be necessary to continue the procedure of testing whether further reallocations would give a better result until the best result is in fact reached.

The best solution is of course recognized by the non-existence of a better one. This is a very satisfactory conclusion since the facilities afforded for making possible the recognition of a best solution also thereby make it possible to arrive at a definite result. The example included above is a simple one and the final result (or an equally good one) could perhaps have been achieved by ordinary trial and error experiments. However, as the numbers of different supply sources and demand categories increase (that is, as the grid takes in more and more squares), the ordinary trial and error experiments would be of little avail amidst the welter of unorganized data. It would never be certain whether an apparently good solution was also the best one obtainable. Linear programming introduces a method of planning so that the experiments may be directed progressively towards the final solution, and it also adds the seal of certainty to the latter when it is reached.

It will always be found in problems of the nature discussed that the number of squares in which allocations are required in the best solution will never exceed the total $(m + n - 1)$ where m represents the number of supply sources and n represents the number of demand categories. This provides some measure of assistance to the research analyst, for if the number of squares used in a possible solution exceeds this total, then it is at once known that a better solution may be found.

The certainty introduced into linear programming relates only to the computations based upon known data. It cannot and does not vouch for the latter. The accuracy of the result depends upon the accuracy of the data. Linear programming cannot produce a result which may be successfully applied to practical problems if the data are faulty. The solutions also imply that the data are fixed. In conditions where, for example, profitability rates fluctuate, the fluctuations may make nonsense of the calculations.

It will be apparent that the larger the number of squares in the grid, so the more lengthy and tiresome would become the process described. There are a number of ways of reducing the work involved. One specific method known as the Simplex Method[1] removes the necessity for the chessboard approach of moving units from one square to another. It is, of course, based upon the same principles but, instead, it sets up algebraic equations and the resolution of these is a process which may be performed on modern computing machines. The latter can take such programmes

1. See S. Vajda, *Introduction to Linear Programming and the Theory of Games*, Methuen.

in their metaphorical stride. This enables many problems to be attempted which would otherwise defy analysis because of the practical difficulties involved. The cost of using a computer may be covered many times by the resultant increases in profit.

It should not be thought that the simplicity of the example quoted in this chapter is representative of linear-programming calculations. Many of these are indeed extremely complex and, as yet more difficult problems are tackled, even more complex methods are developed to deal with them. It is a fact that these methods are not called into use unless the problems are relatively difficult. Perhaps one of the greatest practical difficulties is in recognizing that linear programming is a suitable treatment for a particular problem. Then, once having realized that these methods might be used, the main practical difficulties which face the research worker are involved in the assembling of the necessary data in the form required.

We have referred to the chessboard approach to linear programming in which we move units from one square to another much as we might move chessmen along the rank or file of the board. But the similarity exists only in the fact of movement of units between squares; there is no other similarity to the real movements in the game of chess. Other mathematical models, however, may be used in decision-making which are conceptually based upon the fundamental properties and process of parlour games. Probability models, to which reference has been made throughout the major part of this book, are concerned with chance as a negative type of factor. Chance is the sum of unknown and unknowable factors affecting the outcome of an event and, by implication, such factors are also regarded as uncontrollable. Many decisions, however, must be taken where the chances of success rely not only upon the action of chance factors but also upon the interaction of results obtained by other persons or bodies. Such persons may have interests which are directly opposed to ours. Our success may mean their failure, or vice versa, and in these circumstances we obviously have to take account of their interests. Our results will be affected by theirs, and we can no longer regard blind chance as our only opponent.

Chance, as we understand it, is impartial. Other interested and opposing parties, however, provide a more positive kind of disturbing factor.

It is for circumstances of competition such as these that the theory of games has been developed. Games are used as models to represent the underlying fundamentals of such circumstances, and the theory is intended to deal with competitive situations generally. The aim of a player of games is to gain an advantage over his opponent, so that in a simple zero-sum game[1] an advantage accruing to player A is associated with a disadvantage accruing to player B. Their strategies, or moves in the game, are practical representations of decisions which they make in the course of the game in order to gain the greatest advantage to themselves.

These strategies should be designed so that a player may maximize the number of points he obtains, since that is the object of the game. In practice a player may make his moves as a result of combined guessing and intuitive processes, or he may base them upon a mature consideration of which move out of all possible moves will give him the highest guarantee. These latter moves are based upon a mathematical treatment of all possible outcomes, and it is the strategy of this type with which the theory of games is concerned. The theory demonstrates to each player how he should choose his strategy so as to maximize his advantage.

In an actual game, a player may have to make his moves without foreknowledge of how his opponent will move, but the theory assists him in certain circumstances to make a decision in such a way that, had he known his opponent's intended move, he would still have reached the same decision. The subject is a large one and a great deal of literature is available dealing with its many aspects.[2] Here we can deal only with the fundamental ideas involved in the theory. A simple example will best serve to illustrate these.

Suppose that Mr Black and Mr White play a game in which each can choose one of three numbers. They win or lose points according to the combination of their individual choices. Each

1. See Glossary.

2. e.g., J. D. Williams, *The Compleat Strategyst*, McGraw-Hill.

player then has three choices of action. Suppose also that the points awarded to Black are as follows:

Black's moves	White's moves		
	1.	2.	3.
1	3	−2	−2
2	2	0	2
3	−2	−2	3

The negative values represent losses of points whereas the positive values represent gains of points. Thus if Black selects his first choice when White selects his third choice (that is choice B1 associated with W3) then Black loses 2 points. Black's loss is White's gain, so that the latter gains 2 points. Because Black's loss is always White's gain, and vice versa, it is simpler to consider the results applicable only to one of the players; the results of the game insofar as the other player is concerned are the opposites of the results for the first player. It should be noted that the theory assumes that pay-off results may be measured numerically and that they are known in advance of each game.

In making his choice Black will wish to obtain the highest number of points. This number is 3 which occurs twice in the tabulation, but Black can achieve this total only if White makes move W1 in association with Black's own move B1, or if move W3 is associated with move B3. While Black hopes to obtain these highest points, he must contend with White's attempts to maximize his own points to Black's disadvantage. If Black chooses move B1, he takes the risk that White will not choose W1 for, by taking move W2 or W3, White can make Black lose two points instead of winning 3 points. Black's best move is, in fact, B2 for although he may win nothing he cannot lose anything and may possibly win 2 points.

From White's point of view, W2 is the best move since again, although he may win nothing he cannot lose and may indeed win

2 points. It is thus in their individual interests that Black should select move B2 and that White should select W2. Black chooses the row with the largest minimum (called the *maximin*) as he wishes to ensure that the least he can win is as great as possible and White chooses the column with the smallest maximum (called the *minimax*) since he wishes to ensure that the most he can lose is as small as possible. In effect neither wins or loses any points, but neither Black nor White will regret having taken these moves. Now that Black's move is disclosed as B2, White will realize that had he chosen any move other than W2, he would have lost 2 points. Similarly, given White's move, Black will realize that he would have lost 2 points if he had not chosen B2. Thus, even if Black and White had known what each other's moves would be, they would not have altered their own moves. Therefore, if Black and White always follow the advice offered by the theory, the result would always be the same.

This is a particular case where, in the table of possible outcomes, there is one value which is the maximum in its column and, at the same time, the minimum in its row. This coincident value is said to be positioned at the saddle point of the tabulation or pay-off list; and provided both players direct their strategies towards this point the same result will always be obtained whether or not Black and White are aware of each other's strategies. Each will, however, assume that the other is trying for the result at the saddle point. If either of them fails to make the move indicated by the application of the theory then he will lose points to the other who does follow the theory.

Where there is no saddle-point in a game, it might be thought that Black and White should nevertheless still strive to maximize their relative advantages by reference to the appropriate maximin or minimax by means of simple or pure strategies. A look at the following pay-off (for Black) list should dispel this thought:

	White 1	2
Black 1	4	7
2	6	5

The maximin in any row is 5, and the minimax in any column is 6. Thus Black can guarantee himself a score of at least 5, while White is certain that he will never have to pay more than 6. However, if Black chooses strategy B2, he must take account of the fact that White may discover this before choosing his own strategy. White could then choose W2, thus keeping Black's win down to 5 whereas, had White not obtained prior information as to Black's choice, Black might possibly have won 6. Similarly, with foreknowledge of Black's choice of B1, White can choose W1 and so keep Black's winnings down to 4, whereas otherwise the latter might have won 7.

If, therefore, White knows that Black will always make the same identical move indicated by the position of his maximin, then White does in fact have foreknowledge of Black's intentions and will always be able to ensure that Black does not win his highest possible score. It is thus most important that Black should not let White have this prior knowledge and this in effect means that he cannot always play the same strategy. His only practical alternative, therefore, is to use each of the two strategies open to him. If sometimes he uses strategy B1 and at other times B2, then, for any particular game, White will not know which strategy he has to counter. Black's strategy thus becomes a mixed strategy, and the best way of ensuring that White cannot by any devious ways gain prior information as to his next choice is for Black to let each of his choices be decided by some chance event such as the tossing of a penny or the spinning of a dice. In this way, White will not be able to ferret out the information for the simple reason that Black himself will not know what his move is going to be until the decision is made by the chance event.

Since the majority of games do not have saddle points, the mixed strategy is obviously very important. It should not be thought, however, that this 'decision by chance' removes the skill from the game and that everything is decided by chance alone. It is only the final decision which depends upon chance; the probabilities of B1 and B2 strategies being used will not necessarily be in the ratio $\frac{1}{2}:\frac{1}{2}$. Black can, for example, arrange that the probability of B1 being used is $\frac{3}{4}$ and that the probability of B2 is $\frac{1}{4}$, having regard to the respective pay-offs. Thus Black still controls the overall probabilities of the occurrences of B1 and B2, whereas chance operates only within Black's defined limits to determine whether B1 or B2 should be used in a specific game. Because chance enters into the functioning of a game, it will be seen that the theory extends to the treatment of a series of games rather than to a specific game considered in isolation. A player must sometimes expect to meet bad results as part of a series of results which, taken together, provide the expected average pay-off. Where chance is not involved, the theory gives a direct treatment for each separate game because if both players ideally make their best moves every time then each and every game would be played in an identical manner.

The theory, originally based upon the clear-cut definition of possible outcomes of games, may be applied to other situations where it is possible to compile a pay-off list showing the outcome of all possible decisions. The theory is, however, still in a relatively early stage of its career and there are many problems yet to be overcome. In finite games where the number of possible decisions is fixed, the outcomes can normally be evaluated without much difficulty. Infinite games, as for example where a player has an infinity of possible decisions, are less easily dealt with and, indeed, some of them have no solution in terms of the theory of games. For these and other aspects of the theory, the reader is recommended to refer to *An Introduction to Linear Programming and the Theory of Games* by S. Vajda.[1]

The above outline will serve to give a very general idea of the theory. Strategies are evolved so that, if both players apply the principles of the theory, then neither will regret his decision. This implies that Black and White always make the best possible moves. This may, of course, be crediting either of them with more intelligence than he possesses – always a serious fault in any game or other competitive situation – but whether or not this is so will soon become obvious after only one or two games.

The exposition of the theory in discussing Black's and White's respective strategies may suggest that all that the theory can achieve is a situation of stalemate and that this defeats the object

1. Published by Methuen, 1960.

of the players' intention to win. This situation arises only because it is assumed that both Black and White do in fact follow the theory. A good 'fair' game of skill is one in which two evenly-matched players can, if they make no mistakes, reach a point of checkmate with neither gaining an advantage over the other. If Black defeats White in such a game it is because he is the better player. The theory of games discusses strategies which *should* be followed. If Black and White both adopt their correct strategies, then they will be evenly matched, but as soon as one leaves the paths indicated by the theory he is giving his opponent an opportunity for unexpected success. Players in games are rarely evenly matched. Similarly, not all players will follow the games theory correctly and, in these circumstances, the player who does follow the theory is assured of a successful outcome.

Appendix 1

Coefficient of Correlation

(see page 133).

The formula which defines the coefficient of correlation is

$$r = \frac{\frac{1}{N}\Sigma(x - \bar{x})(y - \bar{y})}{\sigma_x \;\; \sigma_y}$$

where $\bar{x}$ and $\bar{y}$ are the respective arithmetic means of the x values and y values; and where σ_x and σ_y are the standard deviations of the x and y distributions respectively. (The sign Σ indicates that all the separate products $(x - \bar{x})(y - \bar{y})$ for all values of x and y are to be summed.) For the set of values:

(a) $x = 1 \quad y = 2$
(b) $x = 2 \quad y = 4$
(c) $x = 3 \quad y = 6$

the computation would be as follows, tabulated for simplicity.

	x	y	$(x - \bar{x})^2$	$(y - \bar{y})^2$	$(x - \bar{x})(y - \bar{y})$
(a)	1	2	1	4	2
(b)	2	4	0	0	0
(c)	3	6	1	4	2
Totals	6	12	2	8	4
Mean	$\bar{x} = 2$		s.d. = $\frac{2}{3}$		
Mean		$\bar{y} = 4$	s.d. =	$\frac{8}{3}$	

Whence

$$r = \frac{\frac{1}{3}(4)}{(\sqrt{\frac{2}{3}})(\sqrt{\frac{8}{3}})} = \frac{\frac{1}{3}(4)}{\frac{4}{3}} = +1$$

The coefficient may be said to measure how closely the correlation approaches a linear functional relationship; it does not measure the quantitative ratio between x and y. A coefficient value equivalent to unity denotes a perfect functional relationship and all the points representing paired values of x and y would fall on the regression line representing this relationship. For the values of x and y given above, since $r = 1$, then x may always be expressed as a function of y or vice versa. This is confirmed from the separate values; thus $y = 2x$.

The coefficient in fact represents the ratio between two standard deviations. In this ratio the numerator is the standard deviation of the distribution of the calculated values which y would take relative to values of x if all the points fell on the regression line; the denominator is the standard deviation of the distribution of the values of y actually observed.

Appendix 2

The Standard Deviation

The standard deviation is also called the root-mean-square deviation. It is the square root of the mean value of the squares of all the deviations from the distribution mean, and it is calculated as follows:

(i) measure the deviation of each observation from the mean.
(ii) square each deviation and calculate the mean of all the squares.
(iii) extract the square root of this latter mean.

For the set of numbers: 10, 9, 11, 8, and 12

(i) their arithmetic mean is 10, and the respective deviations are:

0 1 1 2 2

(ii) the squares of these deviations are respectively:

0 1 1 4 4

and the mean of these squared numbers is 2.

(iii) the standard deviation is therefore $\sqrt{2}$.

The formula for the standard deviation is thus:

$$\text{s.d.} = \sqrt{\frac{\Sigma(x - \bar{x})^2}{n}}$$

where $\bar{x}$ is the arithmetic mean of a set of n numbers $x_1x_2x_3$ etc. The sign Σ indicates that all the possible values of $(x - \bar{x})^2$ are to be added together.

In the above formula, the summation $\Sigma(x - \bar{x})^2$ represents the total of all the squared deviations; in the example this is equivalent to 10. This total may also be derived as:

$$\sum \bar{x}^2 - \frac{(\Sigma x)^2}{n}$$

For the set of numbers given above:

$$\Sigma x = 10 + 9 + 11 + 8 + 12 = 50$$
$$\Sigma x^2 = 100 + 81 + 121 + 64 + 144 = 510$$

and the sum of the squared deviations is therefore equivalent to:

$$510 - \frac{(50)^2}{5} = 510 - 500 = 10$$

The formula may therefore be expressed differently as:

$$\text{s.d.} = \sqrt{\frac{\sum x^2 - \frac{(\Sigma x)^2}{n}}{n}}$$

While this may look slightly more awkward, it actually provides an easier method of calculation since it obviates the calculating of the individual deviations. It will also give a more accurate result than will the first formula in any instance where the distribution mean cannot be calculated exactly and has to be rounded-off. In such circumstances the rounding-off would mean that each deviation would be slightly inaccurate and that all the inaccuracies would be carried through to the calculations for the standard deviation.

Appendix 3

Least-Squares Method of Fitting a Trend Line to a Distribution

(*see pages 135 and 188*).

If a distribution of the values of y is linear with regard to the values of x (i.e. can be plotted along a straight or linear regression line) then the two quantities x and y may be related to each other in the form

$$y = a + bx$$

where b is the parameter which expresses the gradient of the regression line, and a is the parameter which indicates the value at which the line intersects the y axis.

There are a number of different ways of calculating these parameters; the easiest to explain is one which involves the setting up of two simultaneous equations involving a and b. The values of a and b may then be derived from the solution of these equations. This process is as follows:

(i) substitute in turn each of the paired values of x and y in the equation $y = a + bx$. There will then be as many equations as there are paired values of x and y. Add all these equations to produce a single 'total' equation, which we shall call A.

(ii) Multiply each of the equations built up as in (i) by its respective value of x. Again, add all these equations to produce a single equation, which we call B.

(iii) Solve the simultaneous equations A and B to obtain values of a and b.

For the pairs of values:

x	y
1	2
2	3
3	7

x	y
4	7
5	11

the process would be:

y	x	*Equation as in para (i)*	*Equation as in para (ii)*
		$y = a + bx$	$xy = x(a + bx)$
2	1	$2 = a + b$	$2 = a + b$
3	2	$3 = a + 2b$	$6 = 2a + 4b$
7	3	$7 = a + 3b$	$21 = 3a + 9b$
7	4	$7 = a + 4b$	$28 = 4a + 16b$
11	5	$11 = a + 5b$	$55 = 5a + 25b$
Totals		(*A*) $30 = 5a + 15b$	(*B*) $112 = 15a + 55b$

Solution of the simultaneous equations *A* and *B* produces the values:

$$a = -0{\cdot}6$$
$$b = +2{\cdot}2$$

and the equation for the regression line:

$$y_e = a + bx$$

becomes

$$y_e = 2{\cdot}2(x) - 0{\cdot}6$$

The values plotted along this line are not the values of y actually observed but are instead those which have been calculated to give the line of regression; they are therefore shown as y_e in order to distinguish them from the observed values of y.

The values of y_e (with those of y for comparison) are:

x	y_e	y
1	1·6	2
2	3·8	3
3	6·0	7
4	8·2	7
5	10·4	11

These values are charted in figure 23, and it will be noted that three of the observed values of y are above the regression line and two are below it. Since the regression line is a straight line there is need only to plot the lowest and highest values of y_e; the line connecting these points will automatically pass through all other intermediate values of y_e.

The values of a and b have been calculated from the observed values of x and y. It is important to note that if further values had been available, the results might have been different and it might indeed have been found that subsequent values denied the apparent trend. There is no statistical justification for extending the line of regression beyond the points representing the lowest and highest values of y_e relative to observed values of y. The line may, however, be accepted as an indicator of a trend possibility which, subject to all the difficulties mentioned in Chapters 10 and 13, may assist in the formulation of forecasts for the future.

For a detailed but relatively simple explanation of the basic processes of the least squares method, reference is recommended to M. J. Moroney's *Facts from Figures* (Penguin).

Appendix 4

Geometric Indices and the Time-Reversal Test

The price relatives[1] of individual items will always satisfy the time-reversal test since the same figures are reciprocally involved in the respective calculations:

Price last year: a
,, this year: b

$$\text{Price relative (last year as base)} = \frac{b}{a}$$

$$\text{Price relative (this year as base)} = \frac{a}{b}$$

The product of these two relatives is $\dfrac{b \times a}{a \times b}$ and this is always equal to 1, whatever the values of a and b. Where there are two items involved, the index calculations are:

	First item	*Second item*
Price last year	a	c
,, this year	b	d
Price relative (last year as base)	$\frac{b}{a}$	$\frac{d}{c}$
Price relative (this year as base)	$\frac{a}{b}$	$\frac{c}{d}$

1. See page 152.

The geometric means of these relatives are

(i) last year as base: $\text{g.m.} = \sqrt{\left(\frac{b}{a}\right)\left(\frac{d}{c}\right)}$

(ii) this year as base: $\text{g.m.} = \sqrt{\left(\frac{a}{b}\right)\left(\frac{c}{d}\right)}$

the product of these two geometric means is:

$$\sqrt{\left(\frac{a}{b}\right)\left(\frac{b}{a}\right)\left(\frac{c}{d}\right)\left(\frac{d}{c}\right)}$$

and this again is equal to 1 whatever the values of a, b, c, and d.

Appendix 5

Factorial Designs

A factorial design for experiments is one which arranges that results shall be made available and tabulated in a manner similar to the following tabulation in order that more than one factor may be evaluated within the structure of one series of experiments:

Process	*Operators*			*Total*
	1	2	3	
A	10	12	9	31
B	9	10	8	27
C	9	11	7	27
Total	28	33	24	85

Each process in the example is carried out by each operator, and results are evaluated by reference to some suitable unit of measurement. These measurements are entered in the tabulation. The process results are evaluated in the rows, while the operator results are evaluated in the columns. The totals show the relative ranking scores. For example, Operator 2 obtains consistently better results than those achieved by the others. Process A gives better results than those obtained by the other processes. The results would not usually be as clearly defined as in this example, and in practice their significance would be assessed by analysis of variance techniques. (See Appendix 8.)

Appendix 6

Latin Squares

In mathematics a latin square is defined as a square containing n^2 cells constructed in n rows and n columns, each of the cells containing one of n specified numbers so that every number appears once, and only once, in each row and in each column. A simple square of this type might be as follows, where $n = 3$.

0	1	2
1	2	0
2	0	1

These properties of the latin-square construction may be built into experimental designs so as to remove from the results of the experiment those conflicting variations arising from factors other than the factor which is the immediate subject of the investigation. Thus it may be required to investigate the relative merits of three different processes in the manufacture of a product in circumstances where the observations are to be effected in respect of three machines with three different operators. Here it is necessary to contend not only with the variability arising directly from the use of different processes; there will also intrude a certain amount of variability arising from the fact that three different machines are being used and that three different operators are involved. As an extreme example, it is possible that the most suitable process might in fact appear to be unsuitable on machine 2 merely because that is the oldest machine and is operated by the least efficient operator.

The machine and operator variability may be removed by the use of a latin-square arrangement. Each operator uses each of the processes but on different machines. Similarly each process is carried out on each machine but by different operators. The processes A, B, and C are distributed among the cells in such a way that every process occurs once in each row and once in each column.

Operators \ Machines	1	2	3
1	B	C	A
2	C	A	B
3	A	B	C

In the present arrangement, operator 1 carries out process B on machine 1; whereas operator 2 carries out the same process on machine 3, and operator 3 effects the process on machine 2. This arrangement is just one of twelve possible arrangements each of which would conform to the character of a latin square. The number of possible arrangements increases very rapidly as n^2 (i.e. the number of cells) increases. When $n = 5$, there are 161,280 different squares; when $n = 6$ there are some 373 million. The actual arrangement used in an experiment would be selected by some random method.

The latin-square arrangement shown enables this particular experiment to be effected with only nine observations and it obviates the necessity for each operator to carry out each process on each machine – a necessity which would have involved twenty-seven observations. The evaluation of the separate factors proceeds by the analysis of variance method.

Appendix 7

The Standard Error of the Difference

(between sample means)

It is sometimes desired to test whether two samples are in fact drawn from the same population or whether there is a significant difference between the samples and therefore also between the populations. Measurements made in the course of experiments, for example, will vary slightly within the limits of the calculated value of the standard error of the mean. Such variations may be attributed to chance effects. But measurements revealed in a subsequent experiment may also differ because of the introduction of some new process which in effect has produced a different end-product. The experiment is conducted to test the effect of this new process. How can the research worker tell whether the difference between two experiments (and thus between the samples selected for the experiments) is a chance difference, arising merely from the random selection of samples from the same population, or whether the difference signifies a real difference between the populations represented by the samples?

For this purpose we may calculate the *standard error of the difference*. Just as the standard error of the *mean* is the standard deviation of the distribution of sample means, so the standard error of the *difference* is the standard deviation of the distribution of the differences between sample means. To calculate this latter standard error, we first calculate the *variance* of the distribution of the differences. The variance of a distribution is the mean-square deviation (i.e. the arithmetic mean of the squares of deviations from the distribution mean). It is thus equivalent to the square of the standard deviation (see Appendix 2) and may therefore be expressed as

$$\text{Var} = \sigma^2$$

Similarly, the variance of the distribution of sample means may be expressed:

$$\text{Var}_{\bar{x}} = \frac{\sigma^2}{n}$$

It may also be shown mathematically that the variance of the difference of two independent random variables is equal to the sum of their individual variances. Thus:

$$\text{Var}_{(\bar{x}_1-\bar{x}_2)} = \frac{\sigma_1^2}{n_1} + \frac{\sigma_2^2}{n_2}$$

where the subscripts 1 and 2 indicate the two different samples involved. The standard deviation for the distribution of the differences is equal to the square root of the variance

$$= \sqrt{\frac{\sigma_1^2}{n_1} + \frac{\sigma_2^2}{n_2}}$$

and this is known as the standard error of the difference.

The difference between the two sample means is then measured against the standard error. If the difference between the means exceeds three times the standard error of the difference, such a difference arising only from random sampling variation would, according to the properties of the normal curve, be accounted for only once in about 370 times. Such a difference would therefore be considered statistically significant as indicating that there is a real difference between the populations from which the samples were drawn.

Example

	n	x	σ	σ^2
Sample 1	20,000	49	2·4	5·76
Sample 2	9,000	48	2·7	7·29

For these two samples, the variance of the difference would be

$$\text{Var}_{(\bar{x}_1 - \bar{x}_2)} = \frac{5{\cdot}76}{20{,}000} + \frac{7{\cdot}29}{9{,}000}$$

$$= {\cdot}000288 + {\cdot}00081 = {\cdot}001098$$

The standard error of the difference is therefore equal to

$$\sqrt{{\cdot}001098} = {\cdot}033$$

The difference between $\bar{x}_1$ and $\bar{x}_2$ is 1 (i.e. 49 – 48) and this is about 300 times as great as the standard error of 0·033. The difference is therefore highly significant and it may be accepted that the samples are drawn from different populations.

Appendix 8

Analysis of Variance

The analysis of variance is a technique, originated by Sir R. A. Fisher, which separates the variation ascribable to one group of causes from the variation ascribable to other groups. The variance of a distribution is the square of the standard deviation.

$$\text{Var} = \sigma^2 = \frac{\Sigma(x - \bar{x})^2}{n}$$

A sample variance is likely to understate the variance of the population from which the sample is drawn, particularly if the sample is small, and it is found that a closer estimate of the population variance is obtained by multiplying the sample variance by the factor $\frac{n}{n-1}$. This is known as Bessel's correction and converts the population variance estimate to

$$\frac{\Sigma(x - \bar{x})^2}{n - 1}$$

Before proceeding to the analysis of variance it will be helpful to note that the denominator in this estimate is equivalent to the number of degrees of freedom involved. The number of *degrees of freedom* is the number of values in a set which may be assigned arbitrarily. Thus if $\bar{x}$ is the mean of a set of n observations, then only $(n - 1)$ of the x values may be assigned arbitrarily, for once these have been assigned then, for the given value of $\bar{x}$, the remaining x value is automatically determined and is therefore dependent upon the others. If there are n values in a set, then the total number of degrees of freedom for that set is $(n - 1)$. If this set is composed of m samples each of identical size then as *between the samples* there are $(m - 1)$ degrees of freedom. Again, within each sample there are $\frac{n}{m}$ observations and the number of degrees of freedom within each sample is equal to $\left(\frac{n}{m} - 1\right)$. But there are m samples and therefore the total number of degrees of freedom *within all samples* is $m\left(\frac{n}{m} - 1\right)$ or $(n - m)$.

Thus; if $n = 10$ and $m = 2$, then we have:

	Degrees of freedom	
Between samples	$m - 1$	$= 1$
Within samples	$n - m$	$= 8$
Total	$n - 1$	$= 9$

The analysis of variance then proceeds as follows. If the means of the populations from which m samples are drawn are in fact the same, then the m samples may be treated as independent samples from the same population. The first step, then, is to calculate the means of the individual samples on this assumption. These sample means may then be treated as a sample from the population of means of samples of size $\frac{n}{m}$. This latter population will have a standard deviation equivalent to the standard deviation of the population of individual observations divided by the square root of $\frac{n}{m}$ (see references to the standard error of the mean). Thus

$$\sigma_{\bar{x}} \text{ (= s.d. of means of samples)}$$

$$= \sigma_x \text{ (= s.d. of individual observations)} \div \sqrt{\frac{n}{m}}$$

so that $\sigma_{\bar{x}}^2 \quad = \sigma_x^2 \div \frac{n}{m}$

and $\left(\frac{n}{m}\right) \sigma_{\bar{x}}^2 = \sigma_x^2$

The two sides of this last equation are then computed using different sources of data which are compatible with each other if all the samples are drawn from the same population but not necessarily otherwise. The point of the exercise is that the variance estimates derived are in fact both estimates of the variance of the assumed population from which the samples were drawn. Comparison of the two estimates should therefore give an indication as to whether they may reasonably be accepted as estimates of one value. If they diverge from each other by too great an amount then it may indicate that the samples are actually drawn from different populations. The method employed is best illustrated by an example; this has been kept deliberately simple in order to clarify the principles involved. Consider the following where there are n (= 10) values in m (= 2) samples.

Sample A	*Sample B*
3	1
4	2
5	3
2	3
6	1
20 Sample totals 10	
4 Sample means 2	

First stage: Calculate σ_x^2. This is the *within sample variance*; the variance of observations from their respective sample means and is equivalent to:

$$\frac{\text{(a) sum of all squared deviations}}{\text{(b) number of degrees of freedom}}$$

The value of (a) is derived as the total of the third columns in each of Tables 1 and 2.

TABLE 1
Sample A ($\bar{x} = 4$)

x	*Deviation from* $\bar{x}$	*Square of deviation*
3	−1	1
4	0	0
5	1	1
2	−2	4
6	2	4
20	Totals	10

TABLE 2
Sample B ($\bar{x} = 2$)

x	*Deviation from* x	*Square of deviation*
1	−1	1
2	0	0
3	1	1
3	1	1
1	−1	1
10	Totals	4

The value of (b) has already been shown to be equal to $(n - m) = 8$. Therefore

$$\sigma_x{}^2 = \frac{10 + 4}{8} = 1{\cdot}75$$

Second stage: Calculate $\sigma_{\bar{x}}{}^2$. This is the variance of the population of means of samples and is therefore the *between sample variance.* This is calculated by summing the squares of the deviations of individual sample means from the population (i.e. of samples) mean, and dividing by the number of degrees of freedom. The two sample means are 4 and 2, so that the mean of sample means is 3. The deviations of the sample means from this value are +1 and −1 respectively, so that the sum of the squares of these deviations is 2. The degrees of freedom have already been shown to be equal to $(m - 1) = 1$.

$$\text{Thus } \sigma_{\bar{x}}{}^2 = \frac{2}{1} = 2$$

$$\text{and therefore } \left(\frac{n}{m}\right) \sigma_{\bar{x}}{}^2 = 5 \times 2 = 10$$

Third stage: Summarize the results

Source of variation	*Sums of squares*	*Degrees of freedom*	*Variance estimate*
Between samples	10	1	10
Within samples	14	8	1·75
Totals	24	9	

Fourth stage: Compare the two estimates. This is done by computing the ratio between them, thus

$$\frac{\text{'between' variance}}{\text{'within' variance}}$$

and the value of the ratio, for the given relation between the numbers of degrees of freedom on which the respective variance estimates were based, is checked against tabulated values to test whether the ratio value is significant. This is usually referred to as Snedecor's F test, so named by Snedecor in honour of Sir R. A. Fisher. In the present example the ratio is

$$\frac{10}{1{\cdot}75} = 5{\cdot}7$$

This would be significant at the 5% confidence level but not at the 1% level.[1]

1. D. V. Lindley and J. C. P. Miller, *Cambridge Elementary Statistical Tables*.

Appendix 9

Chi-Square (X^2) Test

This test is derived from the properties of the χ^2 distribution which has many applications in statistics; this test is but one of them. The test provides a technique whereby it is possible to assess the significance of the departure of observed frequencies from the frequencies which would be expected if the data conformed to some theoretical distribution (e.g. the normal or the binomial distribution). It is therefore possible to test the *goodness of fit* – to see how well the distribution of observed data fits the assumed theoretical distribution.

If the distribution of observed data does in fact approximate to an assumed distribution, then we would expect that there should be no significant difference between the *expected* frequencies and the *actual* frequencies. For each actual frequency (A) there will be an expected frequency (E) and the value of χ^2 is calculated as

$$\chi^2 = \sum \frac{(A - E)^2}{E}$$

As a very simple example, let it be supposed that there are two possible and equiprobable outcomes of an event and that the relevant frequencies are:

Outcome	*Expected frequencies* = E	*Actual frequencies* = A
Z	500	600
W	500	400
Total	1,000	1,000

Do these frequencies indicate bias in the results? The first step is to pair off each actual observation with its related expected value nd then, for each pair, to calculate the value

$$\frac{(A - E)^2}{E}$$

and then set up the equation

$$\chi^2 = \sum \frac{(A - E)^2}{E}$$

$$= \frac{(600 - 500)^2}{500} + \frac{(400 - 500)^2}{500} = 40$$

It will be noted that if the expected and actual values were in fact identical the value of (A – E) and therefore also the value of χ^2 would be zero. The calculated value of $\chi^2 = 40$ may therefore be said to measure the extent to which the observed values depart from the expected values. If the value of χ^2 is too great, then the differences between the expected and the actual values cannot be attributed to chance.

Whether or not a calculated value of χ^2 is considered significant is ascertained by reference to tabulated values relative to the relevant number of degrees of freedom involved. In this example there are two possible outcomes; the frequency of one of these may be assigned arbitrarily, for a given total number of events, but the frequency of the second outcome is thereby automatically determined. There is thus only one degree of freedom. Reference to the tables[1] gives the following values of χ^2 at the different levels of confidence.

5% level	$\chi^2 = 3{\cdot}84$
1% "	$= 6{\cdot}63$
0·1% "	$= 10{\cdot}83$

The calculated value of $\chi^2 = 40$ greatly exceeds even the 0·1% level value and, when it is considered that this latter value would

1. D. V. Lindley and J. C. P. Miller, *Cambridge Elementary Statistical Tables.*

be exceeded only once in a thousand times, it becomes evident that the outcomes are not following the pattern described by the binomial distribution. Since the outcomes were stated to be equiprobable, it appears that there is some confounding effect which is disturbing the theoretical equiprobability. In these special circumstances it is evident that bias exists and that inferences from the distribution of observed data as if it conformed to the binomial distribution would not be valid.

In this example the presence of bias could be detected from the original data; but this is a very simple example to illustrate the procedure rather than to solve a particular problem. The χ^2 test may be applied to other distributions where the differences between expected and actual frequencies do not follow an obvious pattern. The test may also be used to detect whether an actual distribution is in fact too good to be true thus pointing to the possibility that the 'observations' may have been tampered with to make them conform.

Readers who are interested in the mathematical justification and further applications of the test are advised to read a textbook on the subject, such as Kendall and Stuart's *Advanced Theory of Statistics* (Griffin).

Glossary of Terms Used

(*See also index for explanations in the text*)

ABSCISSA: (see *Coordinates*)

AXES (X-AXIS AND Y-AXIS): two intersecting lines, perpendicular to each other, by reference to which graphs may be constructed (see figure 23) and along which coordinates are measured.

AXIAL INTERCEPTION, POINT OF: The point at which a curve cuts an axis.

BIAS: systematic distortion of the representativeness of a statistical result as distinct from chance distortions which tend to cancel each other out.

BINOMIAL: a rational integral algebraic expression of two terms, for example $(x + y)$.

BINOMIAL EXPANSION: the representation of a power of a binomial as a sum of individual terms.

CLASS: a subdivision (i.e. a set of values) within a frequency distribution.

CLASS MID-MARK: the mid-value of a class of values.

CONTINUOUS VARIABLE: (see index)

COORDINATES: numbers used to locate a point on a graph relative to the x and y axes, or the lines which measure the perpendicular distances from those axes. The coordinate measured from the y axis parallel to the x axis is called the *abscissa*; and the other is called the *ordinate*.

CORRELATION: interdependence between variables such that when one changes so does the other in some way similar to that denoted by a function but not as explicitly defined. (See Appendix I as to *Correlation coefficient*.)

CYCLE: a periodic movement in a time series.

DEDUCTIVE METHODS: methods of reasoning by the making of inferences from accepted principles.

DEPENDENT VARIABLE: (see *Function*)

DEMOGRAPHIC POPULATION: population of human beings in a community, as of a nation, race, etc., and more specifically as studied for health, disease, and other vital factors.

DISCRETE VARIABLE: (see index)

DISTRIBUTION, FREQUENCY: a set of values of a variable and the frequencies of each value.

EMPIRICAL METHODS: experimental methods not necessarily supported by established theory or law; the accumulation of observational experience rather than the application of logical or mathematical conclusions.

ERROR, STATISTICAL: the difference between an actual observed value of a variable and its 'expected' value (i.e. as derived from some assumed basic law or theory), the deviation arising from some chance effect and not constituting a mathematical mistake. *Sampling error* is the difference between a population parameter and an estimate thereof obtained from a random sample, the difference arising from the fact that only a sample of values has been observed.

EXTRAPOLATION (GRAPH): process of deducing a value greater than or less than all the values charted by a graph on the assumption that a projection of the graph would continue to satisfy the functional relationship demonstrated by the existing graph.

FREQUENCY: the number of times a variable value occurs (e.g. the number of occurrences of an event or measurement or the number of members of a population in specified classes).

FREQUENCY POLYGON: diagrammatic representation of a *frequency distribution* (*q.v.*) of a discrete variable; can also be used for a continuous variable if the frequencies are grouped in classes.

FUNCTION: a quantity (*dependent variable*) which takes on a definite value when a specified value is assigned to another quantity or quantities (*independent variables*); the mathematical expression of the dependent variable in terms of the independent variables so as to identify the relationship between them.

GROSS DOMESTIC PRODUCT: a measure of the value of goods and services produced in the United Kingdom before providing for depreciation or capital consumption.

INDEPENDENT VARIABLE: (see *Function*)

INDUCTIVE METHODS: methods of reasoning by the drawing of conclusions from several known cases.

INTERPOLATION (GRAPH): process of finding a value between two points on a graph by assuming that the point representing the desired value also lies on the line of the graph. The 'reading-off' of a value from a graph.

ONE-TO-ONE CORRESPONDENCE: relationship between two sets of things such that pairs (consisting of one member from each set) may be removed until the sets are simultaneously exhausted.

ORDINATE: (see *Coordinates*)

PARAMETER, POPULATION: a statistical value applicable to a population, as distinct from a *sample statistic* which refers only to a value derived from a sample of the population.

POPULATION, STATISTICAL: any finite or infinite collection of individuals, things, or measurements (actual or potential) defined by some common characteristic; does not refer only to living beings. (See also *Demographic population.*)

RANDOM SELECTION: process of selection of a sample of a population which gives to each member of the population an equal chance of being selected.

RANGE: the difference between the greatest and the least of a set of values.

RECIPROCAL (OF A FRACTION): the fraction formed by interchanging the numerator and the denominator, giving a result equivalent to unity divided by the original fraction.

REGRESSION: a statistical method for investigating relationships between variables by expressing an approximate functional relationship between them.

ROUNDING-OFF: converting a number expressing the value of a quantity to the nearest whole number, or nearest hundred or other

Index

Some other books published by Penguins are described in the following pages

Prelude to Mathematics

W. W. Sawyer

In *Mathematician's Delight*, one of the most popular Pelicans so far published, W. W. Sawyer describes the traditional mathematics of the engineer and scientist. In *Prelude to Mathematics* the emphasis is not on those branches of mathematics which have great practical utility, but on those which are exciting in themselves: mathematics which is strange, novel, apparently impossible; for instance, an arithmetic in which no number is bigger than four. These topics are preceded by an analysis of that enviable attribute 'the mathematical mind'. Professor Sawyer not only shows what mathematicians get out of mathematics, but also what they are trying to do, how mathematics grows, and why there are new mathematical discoveries still to be made. His aim is to give an all-round picture of his subject, and he therefore begins by describing the relationship between pleasure-giving mathematics and that which is the servant of technical and social advance.

Mathematician's Delight

W. W. Sawyer

This volume is designed to convince the general reader that mathematics is not a forbidding science but an attractive mental exercise. Its success in this intention is confirmed by some of the reviews it received on its first appearance:

'It may be recommended with confidence for the light it throws upon the discovery and application of many common mathematical operations' – *The Times Literary Supplement*

'It jumps to life from the start, and sets the reader off with his mind working intelligently and with interest. It relates mathematics to life and thought and points out the value of the practical approach by reminding us that the Pyramids were built on Euclid's principles three thousand years before Euclid thought of them' – *John O' London's*

'The writer clearly not only loves his subject but has unusual gifts as a teacher . . . from start to finish the reader, whose own interests and training may lie in very different fields, can follow the thread' – *Financial News*

Facts from Figures

M. J. Moroney

The enormous success and rapid expansion of statistical techniques in recent years is ample proof of the need for them. The reader will find here a comprehensive introduction to the possibilities of the subject; he is given the how and the why and the wherefore by which he can recognize the kind of problem where Statistics pays dividends. Common sense and simple arithmetic will carry the reader through this book. Every symbol, every principle is explained and illustrated with examples drawn from a wide variety of subjects. The author writes from experience, for he knows the limitations to the usefulness of statistical technique, and appreciates the difficulties of the non-mathematician. The book ranges from purely descriptive statistics, through probability theory, the game of crown and anchor, the design of sampling schemes, production quality control, correlation and ranking methods, to the analysis of variance and covariance.